AF564803

# Theories and Practices of Biotechnology

# Theories and Practices of Biotechnology

P. R. Gilbert

ANMOL PUBLICATIONS PVT. LTD.
NEW DELHI - 110 002 (INDIA)

**ANMOL PUBLICATIONS PVT. LTD.**

**H.O.:** 4374/4B, Ansari Road, Darya Ganj
New Delhi-110 002 (India)
Ph.: 23278000, 23261597
**B.O.:** 1015, Ist Main Road, BSK IIIrd Stage
IIIrd Phase, IIIrd Block
Bangalore - 560 085 (Karnataka)
Ph: 080-41723429, Tel/Fax: 080-2672 3604
Visit us at: www.anmolpublications.com

*Theories and Practics of Biotechnology*



First Published, 2008

ISBN 978-81-261-3272-0

PRINTED IN INDIA

Printed at Mehra Offset Press, Delhi.

# Contents


| | | |
|---|---|---|
| | *Preface* | *vii* |
| | *List of Acronyms* | 1 |
| | *List of Glossary of Terms* | 14 |
| **1.** | Introduction to Theory and Practice of Biotechnology | 95 |
| **2.** | Theories of Biotechnology: An Overviews | 116 |
| **3.** | Practices of Biotechnology: A Modern Perspective | 212 |
| **4.** | Theoretical and Practical Perspectives on Biotechnology | 264 |
| | *Bibliography* | 284 |
| | *Index* | 294 |

# Preface

This book entitled "Theories and Practices of Biotechnology" is designed to acquaint readers with the modern theories and practices of biotechnology at the international level. Over 20 years ago it was discovered that genes, and parts of genes, could be extracted from DNA using protein "scissors" then copied, or cloned. The gene could then be equipped with genetic "switches" to turn it up or down and inserted back into a living organism. This is the basis of genetic modification. It allows the sequences of genes to be determined. Transferring single genes between different plants and animals, turning existing genes up or down, or removing a gene from its original position and placing it in a new position in the same organism are all referred to as *genetic modification (GM)*. The terms genetic engineering (GE) or genetic manipulation also mean the same thing. Plants, animals or microbes which have a new gene inserted into them are called genetically modified organisms (GMOs) or transgenics. The modified gene belongs to the new host but its sequence is altered for example, on the basis of specific information gleaned from studies of other organisms. DNA sequencing is the process of determining the exact order of the chemical building blocks (bases) that make up DNA. It is a laboratory procedure that involves first breaking down the DNA into short pieces, followed by separating the individual fragments using a technique called gel electrophoresis. A bar code pattern of the DNA pieces is produced which can then be read by computer. An enormous volume of information on the genetics of organisms is being generated using this technique and is providing computer experts with a great challenge in handling the data. The Human Genome Project (HGP), which began in 1990, was a 13-year international project involving Australia. Its goal was to discover the approximate 50,000 genes that make up the human genome, make them accessible for further biological study and to determine the complete sequence of the 3 billion DNA subunits. Knowledge gained through this research will allow scientists and doctors of the future to prevent or treat most diseases at the level of genes. While it is now common to screen infants for inherited genetic diseases, further understanding of the human genome will allow doctors to determine a person's chances of developing particular diseases later in life. They will then be able to give advice or treatment to prevent, slow or cure those diseases.

A "clone" is a copy of a plant, animal, microorganism or gene derived from a single common ancestor gene, cell or organism. Identical twins are natural clones from the one fertilized egg. A plant cutting is a clone of the original plant. Some confusion arises because the word *clone* is also applied to genes. A gene is said to be cloned when its sequence is multiplied many times in a common laboratory procedure. The cloning of a gene is a step in determining its sequence and initiating many types of experiments to understand its function and biology. The possibility of human cloning, raised when Scottish scientists at the Roslin Institute created the much-celebrated sheep "Dolly" in 1996, aroused worldwide interest and concern because of its scientific and ethical implications. Injecting a nucleus from a mammary cell into an egg cell without a nucleus and raising an animal from that egg by treating it as a fertilized egg produced dolly. Dolly was therefore a genetic clone of the animal from which the nucleus was taken. The important difference from identical twins is that the donor was a mature sheep and could even be made the mother of the clone. Cloning of animals and plants can be used to develop more efficient ways to produce superior breeds of animals.

Cloning also enables genes to be added (such as those for human proteins) to produce animals that can generate hormones and other pharmaceuticals in milk, eggs or other products. The governments of many nations involved in gene technology have banned the cloning and genetic modification of human individuals. However, cloning technology may soon be available for human benefit to produce whole organs or special tissues from single cells for transplant to humans.

Currently, if someone has an organ transplant, there is quite a high possibility that their body will reject the foreign organ and so they not necessary have to suppress the immune system to lessen the chances of this happening. Stem cells could potentially be used to repair damaged or defective tissues around the body, such as the cells in the pancreas that stop producing insulin in diabetics. The problem with embryonic stem cells is that many people feel that by using a human embryo and then killing it, you are actually killing a potential person. It is for this reason that the United States and other countries are calling for a global ban on all human cloning, including the use of embryonic stem cells. As a result, there has been new research into the potential of what is known as "adult stem cells". More than 30 years ago it was discovered that, in addition to being found in embryos, stem cells are also found in adults. Adult stem cells can be removed from a person without causing any harm. Until very recently, it was thought that adult stem cells were only suitable for cloning a few types of cells and tissues. However, new studies have found adult stem cells with almost the same abilities as embryonic stem cells in various human tissues, including: the spinal cord, the brain, connective tissue and in the blood of the umbilical cord. These studies have shown that it is possible for us to manipulate cells to take on new functions/start doing different jobs simply by placing them in different environments and "reprogramming" them. For example, neural (brain) stem cells in mice have been transformed into the blood stem cells that produce the different types of blood cells. In other animals, bone marrow stem cells have become brain cells and liver cells. In humans, bone marrow stems cells of a patient have been transplanted into the areas of the heart damaged by a heart attack, triggering new growth of heart tissue. Although adult stem cells are already being used successfully to treat a number of diseases, there are a number of problems to be overcome, such as finding and extracting the adult stem cells. Eventually, it may be that, rather than just one type, both types of stem cells will be used for different purposes – whichever proves to be the best for the situation. In the meantime, parallel research continues on both cell types. Although currently the risks of cloning outweigh the possible benefits, there are many different potential uses of human cloning technology.

This book consists of—Introduction to Theory and Practice of Biotechnology; Theories of Biotechnology: An Overviews; Practices of Biotechnology: A Modern Perspective; and Theoretical and Practical Perspectives on Biotechnology etc. besides a large body of list of acronyms, glossary of relevant terms, extensive bibliography, list of websites and links, for further referencing and research.

**—Editor**

## List of Acronyms

| | | |
|---|---|---|
| ABI | — | Applied Biosystems, Inc. |
| ABIC | — | Agricultural Biotechnology Intl. Conf. |
| ACMG | — | American College of Medical Genet. |
| ACS | — | American Chemical Society |
| ADA | — | Americans with Disabilities Act |
| AEC | — | Atomic Energy Commission |
| AES | — | American Electrophoresis Society |
| AFMR | — | American Federation of Medical Research |
| AGSG | — | Alliance of Genetic Support Groups (now Genetic Alliance) |
| AGT | — | Association of Genetic Technologists |
| AHA | — | American Heart Association |
| AIBS | — | Am. Inst. of Biol. Soc. |
| AMIA | — | American Medical Informatics Association |
| ANGIS | — | Australian National Genomic Info. Service |
| ANL | — | Argonne National Laboratory |
| API | — | Application Programming Interface |
| ARP | — | American Registry of Pathology |

| | | |
|---|---|---|
| ASB | — | American Society for Biotechnology |
| ASBMB | — | American Society for Biochemistry & Molecular Biology |
| ASCB | — | American Society for Cell Biology |
| ASCI | — | American Society for Clinical Investigation |
| ASHG | — | American Society for Human Genetics |
| ASIP | — | American Society for Investigative Pathologists |
| ASIS | — | American Society for Information Science |
| ASLME | — | American Society of Law, Medicine, and Ethics |
| ASM | — | American Society for Microbiology |
| ASPET | — | American Society for Pharmacology and Experimental Therapeutics |
| ATCC | — | American Type Culture Collection |
| ATP | — | Advanced Technology Program |
| AVS | — | American Vacuum Society |
| AWCH | — | Adelaide Women and Children's Hospital |
| BAC | — | Bacterial Artificial Chromosome |
| BACR | — | British Association for Cancer Research |
| BCM | — | Baylor College of Medicine |
| BDG | — | Batten's disease gene |
| BER | — | Biological and Environmental Research |
| BNL | — | Brookhaven National Laboratory |

| | | |
|---|---|---|
| BSCS | — | Biological Sciences Curriculum Study |
| BS/SCF | — | Biological Sequence/Structure Computational Facility |
| BTCI | — | BioPharmaceutical Technology Center Institute |
| BTP | — | Biotechnology Training Programs |
| CAE | — | capillary array electrophoresis |
| CaSSS | — | California Separation Science Society |
| CATCMB | — | Center for Advanced Training in Cell and Molecular Biology |
| CCM | — | Chromosome Coordinating Meeting |
| CDC | — | Centers for Disease Control |
| cDNA | — | Complementary DNA |
| CE | — | Capillary Electrophoresis |
| CEPH | — | Centre d'Etude du Polymorphisme Humain |
| CF | — | Cystic Fibrosis |
| CFF | — | Cystic Fibrosis Foundation |
| CHH | — | Cartilage-hair Hypoplasia |
| CHI | — | Cambridge Healthtech Inst. |
| CHOP | — | Children's Hospital of Philadelphia |
| CIMB | — | Center for International Meeting on Biology |
| CIOMS | — | Council for International Organizations of Medical Sciences |
| CIRB | — | Colorado Institute for Research in Biotechnology |

| | | |
|---|---|---|
| CLMA | — | Clinical Laboratory Management Association |
| cM | — | Centimorgan |
| CMT | — | Charcot-Marie-Tooth |
| CONTIG | — | Consortium of Teachers in Genetics |
| CORN | — | Council of Regional Networks for Genetic Services |
| CRADA | — | Cooperative Research and Development Agreement |
| CSHL | — | Cold Spring Harbor Laboratory |
| CTG | — | Trinucleotide Repeat |
| DHHS | — | Dept. of Health and Human Services |
| DIMACS | — | Center for Discrete Math. & Theoretical Comp. Sci. |
| DM | — | Myotonic Dystrophy |
| DMD | — | Duchenne Muscular Dystrophy |
| DNA | — | Deoxyribonucleic acid |
| DOE | — | Department of Energy |
| ECB | — | European Congress on Biotechnology |
| EDS | — | Electronic Data Submission |
| EEOC | — | Equal Employment Opportunity Commission |
| EL.B.A. | — | ELectronics and Biotechnology Advanced |
| EMBL | — | European Mol. Biol. Lab. |
| EMBO | — | European Molecular Biology Organisation |
| EMG | — | Encyclopedia of the Mouse Genome |

| | | |
|---|---|---|
| EMS | — | Environmental Mutagen Society |
| EORTC | — | European Organization for Research and Treatment of Cancer |
| ERDA | — | Energy Research and Development Administration |
| ERI | — | Eleanor Roosevelt Institute |
| ES | — | Embryonic Stem |
| ESF | — | European Science Foundation |
| ESHG | — | European Society of Human Genetics |
| ESI | — | Electrospray Ionization |
| EST | — | Expressed Sequence Tag |
| EUCIB | — | European Collaborative Interspecific Backcross |
| EURESCO | — | European Research Conferences |
| FAP | — | Familial Adenomatous Polyposis |
| FASEB | — | Federation of American Societies for Experimental Biology |
| FCM | — | Flow Cytometry |
| FEBS | — | Fed. of Eur. Biochem. Soc. |
| FMF | — | Familial Mediterranean Fever |
| FRAXA | — | Fragile X locus |
| FSHD | — | Facioscapulohumeral Muscular Dystrophy |
| FTICR | — | Fourier Transform Ioncyclotron Resonance |
| FVEA | — | Fundacion Valenciana de Estudios Avanzados |
| GAS | — | Genome Automation System |

| | | |
|---|---|---|
| GBASE | — | Genome Database of the Mouse |
| GBR | — | Global Business Research |
| GDB | — | Genome Database |
| GDB/OMIM | — | Genome Database/Online Mendelian Inheritance in Man |
| GESTEC | — | Genome Science and Technology Center |
| GIST | — | Genome Informatics System of Transputers |
| GLaRGG | — | Great Lakes Regional Genetics Group |
| GMCRF | — | General Motors Cancer Res. Foundation |
| GMD | — | Genomic Map Design |
| GPI | — | Genetics and Public Issues Program |
| GRAIL | — | Gene Recognition and Analysis Internet Link |
| GRC | — | Gordon Res. Conf. |
| GSA | — | Genetics Societies of America |
| GSDB | — | Genome Sequence Data Base |
| HAEC | — | Human Artificial Episomal Chromosome |
| HELSRD | — | Health Effects and Life Science Research Division |
| HERAC | — | Health and Environmental Research Advisory Committee |
| HGCC | — | Human Genome Coordinating Committee |
| HGM | — | Human Genome Meeting |

| | | |
|---|---|---|
| HGMIS | — | Human Genome Management Information System |
| HGP | — | Human Genome Project |
| HHMI | — | Howard Hughes Medical Institute |
| HICSS | — | Hawaii Intl. Conf. on Systems Sci. |
| HLA | — | Human Leukocyte Antigen |
| HMDP | — | Homology Database |
| HNPCC | — | Hereditary Nonpolyposis Colorectal Cancer |
| HSA | — | Human Serum Albumin |
| HUGO | — | Human Genome Organisation |
| IARC | — | International Agency for Research on Cancer |
| IBC | — | International Business Communications |
| IBEX | — | International Biotechnology EXpo |
| IBF | — | International Business Forum |
| IBI | — | Institute for Biotechnology Information |
| ICES | — | International Council of Electrophoresis Societies |
| ICGEB | — | International Centre for Genetic Engineering and Biotechnology |
| ICHG | — | International Congress of Human Genetics |
| ICPEMC | — | International Commission on Protection Against Environmental Mutagens and Carcinogens |
| ICRF | — | Imperial Cancer Research Fund |

| | | |
|---|---|---|
| IEEE | — | Institute of Electrical and Electronics Engineers |
| IG | — | IntelliGenetics |
| IGES | — | International Genetics Epidemiology Societies |
| IJCAI | — | International Joint Conference on Artificial Intelligence |
| IMA | — | Institute for Mathematics and its Applications |
| IMACS | — | International Association for Mathematics and Computers In Simulation |
| IMAGE | — | Integrated Molecular Analysis of Gene Expression |
| IMGS | — | Intl. Mammalian Genome Society |
| INRIA | — | French Natl. Inst. for Research in Computer Science and Control |
| IOM | — | Institute of Medicine |
| IOR | — | Institute of Religion |
| ISAG | — | Intl. Society for Animal Genetics |
| ISMB | — | Intelligent Systems for Molecular Biology |
| ISONG | — | Intl. Soc. of Nurses in Genet. |
| ISQL | — | Interactive Standard Query Language |
| ISTR | — | Institute for Science Training and Research |
| IUBMB | — | Intl. Union of Biochemistry and Molecular Biology |
| IU | — | Indiana University |

| | | |
|---|---|---|
| IVD | — | *in vitro* Diagnostic |
| JGI | — | Joint Genome Institute |
| LANL | — | Los Alamos National Laboratory |
| LBNL | — | Lawrence Berkeley National Laboratory |
| LLNL | — | Lawrence Livermore National Laboratory |
| LTI | — | Life Technologies, Inc. |
| MALDI | — | matrix-assisted laser desorption ionization |
| MBC | — | Massachusetts Biotechnology Council |
| MBL | — | Marine Biological Laboratory |
| MCD | — | Mouse Cytogenetic Database |
| MDA | — | Muscular Dystrophy Assoc. |
| MEMS | — | Microelectromechanical Systems |
| MGC | — | Mouse Genome Conference |
| MGD | — | Mouse Genome Database |
| MGI | — | Microbial Genome Initiative |
| MHC | — | Major Histocompatibility Complex |
| MIMBD | — | Meet. on the Interconnection of Mol. Biol. Databases |
| MIT | — | Massachusetts Institute of Technology |
| MMRF | — | Marshfield Medical Research Foundation |
| MNBWS | — | Miami Nature Biotechnology Winter Symposium |
| MOD | — | March of Dimes |

| | | |
|---|---|---|
| MOU | — | Memorandum of Understanding |
| MRC | — | Medical Research Council |
| MS | — | Mass Spectrometry |
| NACHGR | — | Natl. Advisory Council for Human Genome Res. |
| NAPBC | — | Natl. Action Plan on Breast Cancer |
| NAS | — | Natl. Academy of Sciences |
| NASA | — | Natl. Aeronautics and Space Administration |
| NBAC | — | Natl. Bioethics Advisory Commission |
| NCGR | — | National Center for Genome Resources |
| NCI | — | Natl. Cancer Institute |
| NCSA | — | National Center for Supercomputing Applications |
| NCSL | — | National Conference of State Legislatures |
| NCSU | — | North Carolina State University |
| NEH | — | National Endowment for the Humanities |
| NFCR | — | National Foundation for Cancer Research |
| NFID | — | National Foundation for Infectious Diseases |
| NHGRI | — | Natl. Human Genome Res. Inst. |
| NICHD | — | National Institute of Child Health and Human Development |
| NIGMS | — | National Institute of General Medical Sciences |

| | | |
|---|---|---|
| NIH | — | National Institutes of Health |
| NIGMS | — | Natl. Institute of General Medical Sciences |
| NIST | — | Natl. Inst. of Standards and Technology |
| NLGLP | — | National Laboratory Gene Library Project |
| NLM | — | National Library of Medicine |
| NMHCC | — | National Managed Health Care Congress |
| NORD | — | National Organization for Rare Disorders |
| NRC | — | National Research Council |
| NRSA | — | National Research Service Award |
| NSF | — | Natl. Sci. Foundation |
| NSGC | — | National Society of Genetic Counselors |
| NSTA | — | National Science Teachers Association |
| NYAS | — | New York Academy of Science |
| OAINN | — | Ohio Aerospace Institute Neural Networks |
| OBER | — | Office of Biological and Environmental Research |
| OHER | — | Office of Health and Environmental Research |
| OMIM | — | Online *Mendelian Inheritance in Man* |
| OPRR | — | Office of Protection from Res. Risks |
| ORAU | — | Oak Ridge Associated Universities |

| | | |
|---|---|---|
| ORISE | — | Oak Ridge Inst. for Science and Education |
| ORF | — | open reading frame |
| ORNL | — | Oak Ridge National Laboratory |
| OSU | — | Oregon State University |
| OTA | — | Office of Technology Assessment |
| PACHG | — | Program Advisory Committee on the Human Genome |
| PCR | — | Polymerase Chain Reaction |
| PFGE | — | Pulsed-field gel electrophresis |
| PG | — | Plant Genome |
| PIR | — | Protein Info. Resource |
| PNNL | — | Pacific Northwest National Laboratory |
| PRIM&R | — | Public Responsibility in Medicine & Research |
| PSC | — | Pittsburgh Supercomputing Center |
| QDFM | — | Quantitative DNA Fiber Mapping |
| RAPD | — | Random Amplified Polymorphic DNA |
| RARA | — | Retinoic Acid Receptor |
| RECOMB | — | Conference on Computational Molecular Biology |
| RFLP | — | Restriction Fragment Length Polymorphism |
| RH | — | Radiation Hybrid |
| RLGS | — | Restriction Landmark Genomic Scanning |
| SASE | — | Sample Sequencing |

| | | |
|---|---|---|
| SBH | — | Sequencing by Hybridization |
| SBIR | — | Small Business Innovation Research |
| SIAM | — | Society for Industrial and Applied Mathematics |
| SIB | — | Society for Industrial Biology |
| SIMS | — | Societal Institute of the Mathematical Sciences |
| SIVB | — | Society for In Vitro Biology |
| SNL | — | Sandia National Laboratory |
| SNP | — | Single Nucleotide Polymorphism |
| SQL | — | Standard Query Language |
| SSLP | — | Single-sequence Length Polymorphism |
| SSR | — | Simple Sequence Repeat |
| STM | — | Scanning Tunneling Microscope |
| STRP | — | Short Tandem Repeat Polymorphism |
| STS | — | Sequence-tagged Site |
| SULS | — | Suffolk Univ. Law School |
| TCR | — | T-cell receptor |
| TIGR/NIST | — | The Inst. for Genomic Res./Natl. Inst. of Standards and Technol. |
| UC | — | University of California |
| UCB | — | University of California, Berkeley |
| UCLA | — | University of California, Los Angeles |
| UICM | — | Univ. of Illinois College of Medicine |

# List of Glossary of Terms

**Amplify:** To increase the number of copies of a DNA sequence, in vivo by inserting into a cloning vector that replicates within a host cell, or in vitro by polymerase chain reaction (PCR).

**Analyte:** A chemical species targeted for qualitative or quantitative analysis.

**Antibiotic:** A class of natural and synthetic compounds that inhibit the growth of or kill other microorganisms.

**Antibody:** An immunoglobulin protein produced by B-lymphocytes of the immune system that binds to a specific antigen molecule. (See monoclonal antibodies, polyclonal antibodies.)

**Anticodon:** A nucleotide base triplet in a transfer RNA molecule that pairs with a complementary base triplet, or codon, in a messenger RNA molecule. See Codon, Messenger RNA, RNA.

**Antigen:** Any foreign substance, such as a virus, bacterium, or protein, that elicits an immune response by stimulating the production of antibodies. (See Antigenic determinant, antigenic switching.)

**Antigenic determinant:** A surface feature of a microorganism or macromolecule, such as a glycoprotein, that elicits an immune response.

**Antigenic switching:** The altering of a microorganism's surface antigens through genetic rearrangement, to elude detection by the host's immune system.

**Antimicrobial agent:** Any chemical or biological agent that harms the growth of microorganisms.

**Antisense DNA:** DNA strand that is complementary (opposite) to the functional gene. Antisense DNA can block the function of the normal sense DNA. This method was used in the construction of the Flavr Savr tomato.

**Antisense RNA:** A complementary RNA sequence that binds to a naturally occurring (sense) mRNA molecule, thus blocking its translation. (See RNA.)

**Aromatic:** A term used to describe cyclic pi-bonded structures of special stability.

**Asexual reproduction:** Nonsexual means of reproduction which can include grafting and budding.

**Autosome:** A chromosome that is not involved in sex determination.

**Bacteriophage (phage or phage particle):** A virus that in- fects bacteria. Altered forms are used as vectors for cloning DNA.

**Bacteriophage (phage):** Any virus that infects bacteria. They were the first organisms used for the study of molecular genetics and are now widely used as cloning vectors. (From Greek *phagein*, to eat.)

**Bacteriorhodopsin:** Pigmented protein found in the plasma membrane of a salt-loving bacterium, *Halobacterium halobium*; it pumps protons out of the cell in response to light.

**Bacteriostat:** A class of antibiotics that prevents growth of bacterial cells.

**Bacterium (plural bacteria):** Common name for any member of the diverse group of procaryotic organisms. Most are single cells, but multicellular forms also exist.

**Bacterium:** A single-celled, microscopic prokaryotic organism: a single cell organism without a distinct nucleus.

**Band:** In gels and blots, bands are visible indications of a particular fragment of a certain size. There may be one to many bands per lane.

**Banoprism:** Scientists at Northwestern University have created a nanoparticle with a new shape that could be a useful tool in the race to detect biological threats. The nanoprism, which resembles a tiny Dorito, exhibits unusual optical properties that could be used to improve biodetectors, allowing them to test for a far greater number of biological warfare agents or diseases at one time.

**Banoscience:** The study of phenomena and manipulation of materials at atomic, molecular and macromolecular scales, where properties differ significantly from those at a larger scale. Nanoscience is primarily the extension of existing sciences into the realms of the extremely small (*nanomaterials, nanochemistry, nanobio, nanophysics*, etc.) while *nanoengineering* represents the extension of the engineering fields into the nano- scale realm (*nanofabrication, nanodevices*, etc.). The exponential growth of nanoscience is largely due to the development of new instruments and related techniques that are used to "routinely" probe and manipulate material at the atomic and molecular level. *Scanning probe microscopies,* analytical electron- beam techniques, epitaxial growth facilities, and *synchrotron* radiation sources are all opening huge opportunities.

**Banoshells:** Procedures that target cancer cells while leaving normal cells untouched, patients controlling the release of medicine in their bodies with an infrared light, and medical test results produced in seconds rather than days – these are three new medical technologies currently being tested by nanotechnology researchers at Rice University. ... The research focuses on nanoshells, a new type of nanoparticle invented by Naomi Halas, Rice professor of electrical and computer engineering and chemistry. Nanoshells are layered nanoparticles whose ability to manipulate light and color can be designed into the nanoparticle by varying the thickness of the nanoparticle's layers.

**Basal body:** Short cylindrical array of microtubules plus their associated proteins found at the base of a eucaryotic cell cilium or flagellum. Serves as a nucleation site for the growth of the axoneme. Closely similar in structure to a centriole.

**Basal lamina (plural basal laminae):** Thin mat of extracellular matrix that separates epithelial sheets, and many types of cells such as muscle cells or fat cells, from connective tissue. Sometimes called a basement membrane.

**Base pair (bp):** A pair of complementary nitrogenous bases in a DNA molecule—adenine-thymine and guanine-cytosine. Also, the unit of measurement for DNA sequences. In DNA, there are four possible bases: cytosine (C), guanine (G), adenine (A), and thymine (T). Cytosine and thymine are pyrimidine bases; adenine and guanine are purine bases. Cytosine is complementary to guanine while adenine is complementary to thymine. If one strand of DNA has the sequence ATTGC then the complementary strand will be TAACG. Two complementary bases constitute a base pair. In RNA, thymine is replaced by uracil.

**Beta-DNA:** The normal form of DNA found in biological systems, which exists as a right-handed helix.

**Beta-Lactamase:** Ampicillin resistance gene. (See Selectable marker.)

**Binding energy:** The reduction in the free energy of a system that occurs when a ligand binds to a receptor. Generally used to describe the total energy required to remove something, or to take a system apart into its constituent particles—for example, to separate two atoms from one another, or to separate an atom into electrons and nuclei.

**Bio-assemblies or Biomolecular Assemblies:** Containing several protein units, DNA loops, lipids, various ligands, etc.

**Bioaugmentation:** Increasing the activity of bacteria that decompose pollutants; a technique used in bioremediation.

**Biochauvinism:** The prejudice that biological systems have an intrinsic superiority that will always give them a monopoly on self-reproduction and intelligence.

**Biocompatible coated materials:** Biocompatible materials usually used in dental and bone implants that enhance biologic fixation, thereby increasing the bond strength between the coated material and bone, and minimize possible biological effects that may result from the implant itself. MeSH, 1999

**Biocompatible materials:** Synthetic or natural materials, other than drugs, that are used to replace or repair any body tissue or bodily function. MeSH, 1973

**Biodefense:** The need for improved national defenses against biological attacks are urgently needed. A real-time reporting system needs to be developed, whereby officials can be informed about an emerging threat before an agent has a chance to affect thousands of people. The development of integrated systems for detecting and monitoring biological agents is instrumental to this goal. The development of vaccines and novel detection platforms will enable preparedness, while also benefiting human health.

**Biodiversity:** The wide diversity and interrelatedness of earth organisms based on genetic and environmental factors.

**Biodynotics Biologically Inspired Multifunctional Dynamic Robotics:** A multidisciplinary, multi-pronged approach with far reaching impact on robotic capabilities for national security applications. Biologically Inspired Multifunctional Dynamic Robotics: BIODYNOTICS will explore the following areas: DYNAMIC MOBILITY, BEHAVIOR, INTEGRATION Biological Sciences, Defense Sciences Office, DARPA.

**Bioengineering:** Is rooted in physics, mathematics, chemistry, biology, and the life sciences. It is the application of a systematic, quantitative, and integrative way of thinking about and approaching the solutions of problems important to biology, medical research, clinical proactive, and population studies. The NIH Bioengineering Consortium agreed on the following definition for bioengineering research on biology, medicine, behavior, or health recognizing that no definition could completely eliminate overlap with other research disciplines or preclude variations in interpretation by different individuals and organizations. Integrates physical, chemical, or mathematical sciences and engineering principles for the study of biology, medicine, behavior, or health. It advances fundamental concepts, creates knowledge for the molecular to the organ systems levels, and develops innovative biologics, materials, processes, implants, devices, and informatics approaches for the prevention, diagnosis, and treatment of disease, for patient rehabilitation, and for improving health.

**Bioenrichment:** Adding nutrients or oxygen to increase microbial breakdown of pollutants.

**Biofabrication:** Includes nanoparticle delivery systems, biomaterials, tissue engineering, implants and prostheses. Using biological processes to synthesize and manufacture chemicals and materials of high value to the Department of Defense. Biological processes are characterized by: low energy barriers (~10Kcal, high temperatures/pressures not required); high reaction-, regio- and stereo- specificity (protection/ deprotection wastes and catalyst poisoning avoided); spatio-temporal control of materials synthesis of defined composition and size with angstrom- level precision; and local control of the dielectric environment (largely eliminating the need for toxic solvents). Potential target materials and chemicals include composites with enhanced mechanical properties, ultra-low k dielectrics and thermoelectrics, optoelectronic materials, photonic devices (waveguides and logic elements), electronic materials (e.g., GaN, InGaN, AlGaN), elastomers, energetic materials, and adhesives. Biofabrication, DARPA.

**Biologics:** Agents, such as vaccines, that give immunity to diseases or harmful biotic stresses.

**Biomass:** The total dry weight of all organisms in a particular sample, population, or area.

**Biomaterials:** Synthetic or natural materials that can replace or augment tissues, organs or body functions.

**Biomechanics:** Mechanical structures of living organisms (especially muscles and bones).

**Biomedical Nanotechnology:** (1) the comprehensive monitoring, control, construction, repair, defense, and improvement of all human biological systems, working from the molecular level, using engineered nanodevices and nanostructures; (2) the science and technology of diagnosing, treating, and preventing disease and traumatic injury, of relieving pain, and of preserving and improving human health, using molecular tools and molecular knowledge of the human body; (3) the employment of molecular machine systems to address medical problems, using molecular knowledge to maintain and improve human health at the molecular scale. (4) A rapidly expanding field that includes many potential technologies and approaches. The key to this definition is that phenomena and materials at the nanometer scale are known to have properties that are uniquely attributable to that scale length. Nanomedicine could similarly be defined [as nanotechnology] as the design, synthesis, or application of materials, devices, or technologies in the nanometer-scale for the basic understanding, diagnosis, and / or treatment of disease. Canadian Institute of Health Research, Regenerative Medicine and Nanomedicine, RFA, 2003. The monitoring, repair, construction and control of human biological systems at the molecular level, using engineered nanodevices and nanostructures. (5) the comprehensive monitoring, control, construction, repair, defense, and improvement of all human biological systems, working from the molecular level, using engineered nanodevices and nanostructures; (6) the science and technology of diagnosing, treating, and preventing disease and traumatic injury, of relieving pain, and of preserving and improving human health, using molecular tools and molecular knowledge of the human body; (7) the employment of molecular machine systems to address medical problems, using molecular knowledge to maintain and improve human health at the molecular scale. The use of molecular-scale devices (nanotechnology) to repair damage and boost the immune system. Vapour grown carbon fibres are obtained as shown in this SEM image. The diameters of these fibres can vary from 100 nm to 500 nm. As part of the National Institutes of Health (NIH) Roadmap for Medical Research [nihroadmap.nih. gov], the NIH [nih.gov] will establish a handful of nanomedicine centers. These centers will be staffed by a highly interdisciplinary scientific crew including biologists, physicians, mathematicians, engineers and computer scientists. Research conducted over the first few years will be spent gathering extensive information about how molecular machines are built. A key activity during this time will be the development of a new kind of vocabulary, or lexicon, to define biological parts and processes in engineering terms. Once researchers have completely catalogued the interactions between and within molecules, they can begin to look for patterns and a higher order of connectedness than is possible to identify with current experimental methods. Mapping these networks and understanding how they change over time will be a crucial step toward helping scientists understand nature's rules of biological design. Understanding these rules will, in many years' time, enable researchers to use this information to address biological issues in unhealthy cells.

**Biomedical polymers**: More and more therapeutic problems are relevant to the use of polymer- based therapeutic aids for a limited period of time, namely the healing time related to the outstanding capacity of living systems to self- repair ... After healing the remaining prosthetic materials or devices become foreign residues or wastes that have to be eliminated from the body. Nowadays, biocompatible polymers that can degrade in the body are developed. The degradation and the elimination of degradation by-products depend on rather complex phenomena that are presently reflected inconsistently by terms issued from the tradition because each domain has developed its own terminology almost independently. This is a source of misunderstandings, confusions and misperceptions among scientists, surgeons, pharmacists, journalists and politicians, the situation being increased by the introduction of degradable polymers in plastic waste management and environmental protection. Therefore, it is urgent to reflect the various phenomena by specific terms, harmonize and enforce their use by the people active in the biomedical, pharmacological and environmental fields, and, last but not least by the publishing media and journalists. IUPAC, Terminology for biomedical (therapeutic) polymers.

**BioMEMS Biological MicroElectro Mechanical Systems:** Highlights the technical advances in the field that are leading a revolution in medicine, and creating a new generation of analytical devices for medical diagnostics. The meeting will encompass technology developments in micro & nano drug delivery, interface of nanotech and tissue engineering, microfluidics, and miniaturized total analysis systems (microTAS), biosensors, innovations in mass spec, and nanoscale imaging *BioMEMS and Nanotech World*, Aug. 16- 17, 2004, Washington DC

**BioMEMS:** MEMS used in medicine, that use microchips.

**Biomimetic Chemistry:** Knowledge of biochemistry, analytical chemistry, polymer science, and biomimetic chemistry is linked and applied to research in designing new molecules, molecular assemblies, and macromolecules having biomimetic functions. These new bio-related materials of high performance, including, for example, enzyme models, synthetic cell membranes, and biodegradable polymers, are prepared, tested, and constantly improved in this division for industrial scale production.

**Biomimetic materials:** Materials fabricated by BIOMIMETICS techniques, i.e., based on natural processes found in biological systems. [MeSH 2003] (2) Materials that imitate, copy, or learn from nature.

**Biomimetic:** Imitating, copying, or learning from nature. Nanotechnology already exists in nature; thus, nanoscientists have a wide variety of components and tricks already available. An interdisciplinary field in materials science, ENGINEERING, and BIOLOGY, studying the use of biological principles for synthesis or fabrication of BIOMIMETIC MATERIALS. There is a need to develop the next generation of restorative materials and medical implants. New avenues of scientific inquiry may enable the development of biomaterials that are safe, reliable, "smart", long-lasting, and perform ideally in their respective biological environments.... Over the last few years biomimetics and tissue engineering have emerged as a new vision in the field of tissue and organ repair and restoration. Biomimetics and tissue engineering are interdisciplinary fields that combine information from the study of biological structures and their functions with physics, mathematics, chemistry and engineering for the generation of new materials, tissues and organs. In the area of craniofacial, oral and dental principles from biomimetics and tissue engineering are applied to developing dental and facial implants, new polymers for guided tissue regeneration used in treating periodontal disease and bone and connective tissue defects, coral- based hydroxyapatite replicas for reconstruction of alveolar ridges and other osseous defects, temporomandibular joint (TMJ) and other joint prostheses, formation of bone matrix substitutes, and artificial replicas of bone, skin, and mucosa. The term biomimetics was coined in 1972 in the context of artificial enzymes; it might be defined broadly as "the abstraction of good design from nature". It is a fact of everyday life that nature has managed to built materials and 'devices' with breathtaking functionality, heterogeneity and stability by using a comparatively limited number of building blocks (the whole range of synthetic materials is restricted to man- made engineering). The basic concepts of nature are often simple; it is the way in which building blocks and materials are arranged that results in functionality. Among the most simple and abundant themes of nature is self- assembly: lipids assemble in sheets to form cell membranes, proteins assemble into functional enzymes, cellular 'sensors', fibers, or virus coats, and DNA assembles in double strands to provide the very basis for live: replication. Biomimetics Study of the structure and function of biological substances to make artificial products that mimic the natural ones.

**BIOML Biopolymer Markup Language:** Designed by the BIOML core team at Proteometrics, LLC and Proteometrics Canada Ltd. It is to be used "for the annotation of biopolymer sequence information. BIOML allows the full specification of all experimental information known about molecular entities composed of biopolymers, for example, proteins and genes."

**Biomolecular materials:** An emerging discipline, materials whose properties are abstracted from biology. They share many of the characteristics of biological materials but are not necessarily of biological origin. For example, they may be inorganic materials that are organized or processed in a biomimetic fashion. A key feature of biological and biomolecular materials is their ability to undergo *self- assembly.*

**Biomolecular Nanotechnology:** Nanotechnology existing in living systems and resulting from our ability to use biomolecules as components for molecular nanotechnology.

**Biomotors:** Driven by energy sources such as *adenosine triphosphate (ATP)* for chemical transduction and other processes. These biomotors are considered to be biomolecular and are discussed in the body of this report, but strictly speaking they do not conform to the panel's definition of self- assembly.

**Bionanotechnology:** Includes molecular motors, biomaterials, single molecule manipulation technologies, biochip technologies, etc.

**Biopolymeroptoelectromechanical Systems [BioPOEMS]:** Combining optics and microelectromechanical systems, and used in biological applications.

**Biopolymers:** Macromolecules (including proteins, nucleic acids and polysaccharides) formed by living organisms.

**Bioremediation:** The use of microorganisms to remedy environmental problems. See Bioaugmentation, Bioenrichment.

**Biorobotics:** Our research focuses on the role of sensing and mechanical design in motor control, in both robots and humans. This work draws upon diverse disciplines, including biomechanics, systems analysis, and neurophysiology. The main approach is experimental, although analysis and simulation play important parts. In conjunction with industrial partners, we are developing applications of this research in biomedical instrumentation, teleoperated robots, and intelligent sensors.

**Biosensor:** The term "biosensor" is a general designation that denotes either a sensor to detect a biological substance or a sensor which incorporates the use of biological molecules such as antibodies or enzymes. Biosensors are a subcategory of chemical sensors.

**Biostasis:** A condition in which an organism's cell and tissue structure are preserved, allowing later restoration by cell repair machines. Applicable to cryonics.

**Biotechnology:** The scientific manipulation of living organisms, especially at the molecular genetic level, to produce useful products. Gene splicing and use of recombinant DNA (rDNA) are major techniques used. Technology for working with biological systems. Includes genetic engineering, human and veterinary medicine, crop and animal breeding, diagnostics, pharmaceuticals, forensics, etc. Narrow sense: Genetic engineering.

**Biotic stress:** Living organisms which can harm plants , such as viruses, fungi, and bacteria, and harmful insects. See Abiotic stress.

**Biovorous:** From "biovore;" an organism capable of converting biological material into energy for sustenance.

**Bipolar-junction transistor:** Transistor with n-type and p-type semiconductors having base-emitter and collector-base junctions.

**Blocks Substitution Matrix:** A substitution matrix in which scores for each position are derived from *observations* of the frequencies of substitutions in blocks of local alignments in related proteins. Each matrix is tailored to a particular evolutionary distance. In the BLOSUM62 matrix, for example, the alignment from which scores were derived was created using sequences sharing no more than 62% identity. Sequences more identical than 62% are represented by a single sequence in the alignment so as to avoid over-weighting closely related family members. (Henikoff and Henikoff)

**Blot, northern:** A pattern of RNA fragments transferred to a nitrocellulose membrane from a gel. The gel has undergone electrophoresis to separate fragments according to size. The RNA fragments are arranged in different lanes for each sample. Each lane contains bands which are fragments of different sizes. Radioactive probes are often used to visualized particular bands by autoradiography.

**Blot, southern:** A pattern of DNA fragments transferred to a nitrocellulose membrane from a gel. The gel has undergone electrophoresis to separate fragments according to size. The DNA fragments are arranged in different lanes for each sample. Each lane contains bands which are fragments of different sizes. Radioactive probes are often used to visualized particular bands by autoradiography.

**Blot, western:** A pattern of proteins transferred to a nitrocellulose membrane from a gel. The gel has undergone electrophoresis to separate fragments according to size. The proteins are arranged in different lanes for each sample. Each lane contains bands which are fragments of different sizes. Labeled antibody probes are often used to visualize particular bands.

**Bone substitutes:** Synthetic or natural materials for the replacement of bones or bone tissue. They include hard tissue replacement polymers, natural coral, hydroxyapatite, beta- tricalcium phosphate, and various other biomaterials. The bone substitutes as inert materials can be incorporated into surrounding tissue or gradually replaced by original tissue.

**Brownian Assembly:** Brownian motion in a fluid brings molecules together in various position and orientations. If molecules have suitable complementary surfaces, they can bind, assembling to form a specific structure. Brownian assembly is a less paradoxical name for self-assembly (how can a structure assemble itself, or do anything, when it does not yet exist?).

**Bt (*Bacillus thuringiensis*):** A soil bacterium that produces insecticidal proteins. There are several different kinds of proteins produced by different strains of Bt. Some are effective against larvae of moths and butterflies. Others are effective against larvae of beetles. The Bt protein has been introduced into various crops as a built-in insecticide.

**Calorie:** Unit of heat. One calorie (small "c") is the amount of heat needed to raise the temperature of 1 gram of water by 1 °C. A kilocalorie is the unit used to describe the energy content of foods.

**CAP (catabolite gene activator protein):** Gene regulatory protein in procaryotes that, when glucose is absent, activates genes responsible for the breakdown of alternative carbon sources.

**Capacitor:** Energy storage circuit element having two conductors separated by an insulator.

**Carcinogen:** A substance that induces cancer.

**Carcinoma:** A malignant tumor derived from epithelial tissue, which forms the skin and outer cell layers of internal organs.

**Catalyst:** A substance that promotes a chemical reaction by lowering the activation energy of a chemical reaction, but which itself remains unaltered at the end of the reaction. (See Catalytic antibody, Catalytic RNA.)

**Catalytic antibody (abzyme):** An antibody selected for its ability to catalyze a chemical reaction by binding to and stabilizing the transition state intermediate.

**Catalytic RNA (ribozyme):** A natural or synthetic RNA molecule that cuts an RNA substrate.

**Cation:** A positively charged ion.

**cDNA library:** A library composed of complementary copies of cellular mRNAs. (See Library.)

**cDNA:** DNA synthesized from an RNA template using reverse transcriptase. Complementary DNA to a particular RNA fragment.

**Cell pharmacology:** Delivery of drugs by medical nanomachines to exact locations in the body.

**Cell Repair Machine:** A system including nanocomputers and molecular scale sensors and tools, programmed to repair damage to cells and tissues. Molecular and nanoscale machines with sensors, nanocomputers and tools, programmed to detect and repair damage to cells and tissues, which could even report back to and receive instructions from a human doctor if needed.

**Cell surgery:** In medical nanorobotics, modifying cellular structures using medical nanomachines. Modifying cellular structures using medical nanomachines.

**Cell:** A membrane-bound unit, typically microns in diameter. All plants and animals are made up of one or more cells (trillions, in the case of human beings). In general, each cell of a multicellular organism contains a nucleus holding all of the genetic information of the organism. (2) A small structural unit, surrounded by a membrane, making up living things.

**Cellular oncogene (proto-oncogene):** A normal gene that when mutated or improperly expressed contributes to the development of cancer. (See Oncogene.)

**Centers of origin:** Usually the location in the world where the oldest cultivation of a particular crop has been identified.

**Central dogma:** Francis Crick's seminal concept that in nature genetic information generally flows from DNA to RNA to protein.

**Centrifugation:** Separating molecules by size or density using centrifugal forces generated by a spinning rotor. G forces of several hundred thousand times gravity are generated in ultracentrifugation. (See Density gradient centrifugation.)

**Centromere:** The central portion of the chromosome to which the spindle fibers attach during mitotic and meiotic division.

**Chemotherapy:** A treatment for cancers that involves administering chemicals toxic to malignant cells.

**Chloramphenicol:** An antibiotic that interferes with protein synthesis.

**Chromatid:** Each of the two daughter strands of a duplicated chromosome joined at the centromere during mitosis and meiosis.

**Chromosome walking:** Working from a flanking DNA marker, overlapping clones are successively identified that span a chromosomal region of interest. (See Chromosome.)

**Chromosome:** A single DNA molecule, a tightly coiled strant of DNA, condensed into a compact structure in vivo by complexing with accessory histones or histone-like proteins. Chromosomes exist in pairs in higher eukaryotes. (See Chromosome walking.) A linear structure in the nucleus of plants and animals that is visible in light microscopy when stained. The chromosome is a single, long, linear molecule of DNA and associated proteins. Bacteria have a single circular chromosome; other organisms may have many linear chromosomes.

**Cistron:** A DNA sequence that codes for a specific polypeptide; a gene. See DNA, Gene.

**Clone:** An exact genetic replica of a specific gene or an entire organism. See Cloning.

**Cloning vector:** A DNA molecule capable of autonomous replication within the cloning host cell (e.g *E. coli*). The vectors contain restriction enzyme sites for insertion of foreign DNA. Cloning vectors are derived from bacterial plasmids, bacteriophages, or viruses.

**Cloning:** The mitotic division of a progenitor cell to give rise to a population of identical daughter cells or clones. (See Directional cloning, Megabase cloning, Molecular cloning, Subcloning.)

**Coat protein (capsid):** The coating of a protein that enclosed the nucleic acid core of a virus.

**Codon:** A group of three nucleotides that specifies addition of one of the 20 amino acids during translation of an mRNA into a polypeptide. Strings of codons form genes and strings of genes form chromosomes. Set of three nucleotides that specify a particular amino acid during protein synthesis.

**Coenzyme (cofactor):** An organic molecule, such as a vitamin, that binds to an enzyme and is required for its catalytic activity.

**Colony:** A group of identical cells (clones) derived from a single progenitor cell.

**Combinatorial materials design:** Uses computing power (sometimes together with massive parallel experimentation) to screen many different materials possibilities to optimize properties for specific applications (e.g., catalysts, drugs, optical materials).

**Commensalism:** The close association of two or more dissimilar organisms where the association is advantageous to one and doesn't affect the other(s). See Parasitism, Symbiosis.

**Competency:** An ephemeral state, induced by treatment with cold cations, during which bacterial cells are capable of uptaking foreign DNA.

**Complementary DNA or RNA:** The matching strand of a DNA or RNA molecule to which its bases pair.

**Complementary nucleotides:** Members of the pairs adenine-thymine, adenine-uracil, and guaninecytosine that have the ability to hydrogen bond to one another. (See nucleotide.)

**Concatemer:** A DNA segment composed of repeated sequences linked end to end.

**Configuration space:** A mathematical space describing the three-dimensional configuration of a system of particles (e.g., atoms in a nanomechanical structure) as a single point; the configuration space for an *N* particle system has 3*N* dimensions.

**Conformation:** The shape of a molecule. A molecular geometry that differs from other geometries chiefly by rotation about single or triple bonds; distinct conformations (termed con formers) are associated with distinct potential wells. Typical biomolecules and products of organic synthesis can interconvert among many conformations; typical diamondoid structures are locked into a single potential well, and thus lack conformational flexibility.

**Conjugation:** The joining of two bacteria cells when genetic material is transferred from one bacterium to another.

**Constitutive promoter:** An unregulated promoter that allows for continual transcription of its associated gene. (See Promoter.)

**Contiguous (contig) map:** The alignment of sequence data from large, adjacent regions of the genome to produce a continuous nucleotide sequence across a chromosomal region.

**Coulomb [C]:** Measure of electrical charge: 1 C is an amount of charge equal to that of about $6.24 \times 10^{18}$ electrons.

**Coupling:** A connection between more than one object or energy pathway so that together they function as a single unit.

**Cross-hybridization:** The hydrogen bonding of a single-stranded DNA sequence that is partially but not entirely complementary to a singlestranded substrate. Often, this involves hybridizing a DNA probe for a specific DNA sequence to the homologous sequences of different species.

**Crossing-over:** The exchange of DNA sequences between chromatids of homologous chromosomes during meiosis.

**Cross-pollination:** Fertilization of a plant from a plant with a different genetic makeup.

**Crosstalk:** Electromagnetic noise transmitted between leads or circuits in close proximity to each other.

**Crybiology:** The science of biology at low temperatures; research in cryobiology has made possible the freezing and storing of sperm and blood for later use.

**Crystal Lattice:** The regular three-dimensional pattern of atoms in a crystal.

**Crystal structure:** For crystalline materials, the manner in which atoms or ions are arrayed in space. It is defined in terms of the unit cell geometry and the atom positions within the unit cell.

**Crystallescence:** In medical nanorobotics, the crystallization of solid solute that is offloaded by nanorobot sorting rotors at a concentration that exceeds the solvation capacity of the surrounding solvent.

**Culture:** A particular kind of organism growing in a laboratory medium.

**Curie temperature (also Curie point) ($T_c$):** The temperature above which a ferromagnetic or ferrimagnetic material becomes paramagnetic. For iron the Curie point is 760oC and for nickel 356°C.

**Current [A]:** Measure of rate of flow of electric charge: a one-ampere current is a flow of 1 C of charge per second.

**Cutoff:** Condition in a diode or bipolar junction transistor in which the potential across a p-n junction prevents current flow.

**Cyclic AMP (cyclic adenosine monophosphate):** A second messenger that regulates many intracellular reactions by transducing signals from extracellular growth factors to cellular metabolic pathways.

**Cyclic:** A structure is termed cyclic if its covalent bonds form one or more rings.

**Cytogenetics:** Study that relates the appearance and behavior of chromosomes to genetic phenomenon.

**Dalton:** A unit of measurement equal to the mass of a hydrogen atom, 1.67 x 10E-24 gram/L (Avogadro's number).

**Denature:** To induce structural alterations that disrupt the biological activity of a molecule. Often refers to breaking hydrogen bonds between base pairs in double-stranded nucleic acid molecules to produce in single-stranded polynucleotides or altering the secondary and tertiary structure of a protein, destroying its activity.

**Density gradient centrifugation:** High-speed centrifugation in which molecules "float" at a point where their density equals that in a gradient of cesium chloride or sucrose. (See Centrifugation.)

**Diabetes:** A disease associated with the absence or reduced levels of insulin, a hormone essential for the transport of glucose to cells.

**Dideoxynucleotide (didN):** A deoxynucleotide that lacks a 3' hydroxyl group, and is thus unable to form a 3'-5' phosphodiester bond necessary for chain elongation. Dideoxynucleotides are used in DNA sequencing and the treatment of viral diseases. (See Nucleotide.)

**Digest:** To cut DNA molecules with one or more restriction endonucleases.

**Diploid cell:** A cell which contains two copies of each chromosome. See Haploid cell.

**Directional cloning:** DNA insert and vector molecules are digested with two different restriction enzymes to create noncomplementary sticky ends at either end of each restriction fragment. This allows the insert to be ligated to the vector in a specific orientation and prevents the vector from recircularizing. (See Cloning.)

**DNA:** DNA is the recipe for life. DNA is a molecule found in the nucleus of every cell and is made up of 4 nucleotides Adenine (A), Guanine (G), Thymine (T), and Cystosine (C). The order of the nucleotides in the DNA strand holds a code of information for the cell.

**DNA (Deoxyribonucleic acid):** An organic acid and polymer composed of four nitrogenous bases—adenine, thymine, cytosine, and guanine linked via intervening units of phosphate and the pentose sugar deoxyribose. DNA is the genetic material of most organisms and usually exists as a double-stranded molecule in which two antiparallel strands are held together by hydrogen bonds between adeninethymine and cytosine-guanine. (See b-DNA, cDNA, Complementary DNA or RNA, DNA polymorphism, DNA sequencing, Double-stranded complementary DNA, Duplex DNA, Z-DNA.)

**DNA (Deoxyriobonucleic Acid):** DNA molecules are long chains consisting of four kinds of nucleotides; the order of these nucleotides encodes the information needed to construct protein molecules. These in turn make up much of the molecular machinery of the cell. DNA is the genetic material of cells.

**DNA Chip:** Also: Gene Chip and DNA Microchip. A purpose built microchip used to identify mutations or alterations in a gene's DNA.

**DNA diagnosis:** The use of DNA polymorphisms to detect the presence of a disease gene.

**DNA diagnostics - miniaturization of:** In the areas of sample preparation and assay, it is clear that miniaturization is key. To reduce the size of samples by a factor of 10 or greater, barriers in microfluidics, micromachining, robotics, microchemistry, nucleic acid chemistry, and surface chemistry must be overcome. To implement miniaturized protocols accurately and efficiently, substantial automation of the process will be required. In the development of miniaturized systems, it is essential that the system can be adapted for high levels of parallelization. Miniaturization poses significant technological risks. Currently, there exists no universally accepted precedent for the handling, replication, amplification, or cloning of DNA in nanoliter volumes. Due to the size and charge of the DNA molecule, and the relative instability of many of the enzymes involved in the sample preparation processes, nanoliter and less volumes may pose substantial challenges. In addition, interactions of the biological molecules with the surfaces of the reaction chambers must be minimized. For some methodologies, it is not clear what the optimal sample will be, so substantial improvement in DNA fragmentation technologies or DNA cloning vectors may be required for the ultimate efficient application to diagnostics. Improvements in any of these areas are likely to be of value to other non-DNA based diagnostic applications such as antibody screening protocols and enzyme based diagnostics, because miniaturized robotic or micro- electro mechanical systems developed for DNA could be modified to be used for these purposes.

**DNA fingerprint:** The unique pattern of DNA fragments identified by Southern hybridization (using a probe that binds to a polymorphic region of DNA) or by polymerase chain reaction (using primers flanking the polymorphic region).

**DNA polymorphism:** One of two or more alternate forms (alleles) of a chromosomal locus that differ in nucleotide sequence or have variable numbers of repeated nucleotide units. (See Allele.)

**DNA Probes:** A DNA or nucleic acid probe is a short strand of DNA that locates and binds to its complementary sequence in samples containing single strands of DNA or RNA enabling identification of specific sequences. Nucleic acid probe assays exploit the fundamental hybridization reaction that occurs spontaneously between two complementary DNA:DNA or DNA:RNA strands. As in immunoassays, detection of the hybrid requires that the probe be labeled. Various direct and indirect methods have been devised for the detection of the hybrid. Direct labeling involves attaching the label directly to the probe sequence; indirect labeling binds an antibody to the DNA:DNA or DNA:RNA hybrid. As in immunoassays, non-isotopically-labeled probes are preferred over radio-labeled probes primarily because of radiation hazards, disposal problems, and short reagent shelf life. In addition, the factors determining the detection limits of hybridization assays based on labeled probes are similar to those in immunoassays. Therefore, the development of a simple, inexpensive and sensitive direct detection system which eliminates the use of labels is highly desirable.

**DNA Sequencing:** There are two main classical methods for sequencing DNA: The first method, developed by Allan Maxam and Walter Gilbert, involves chemicals used to cleave the DNA at certain positions, generating a set of fragments that differ by one nucleotide. The second method, developed by Fred Sanger and Alan Coulson, involves enzymatic synthesis of DNA strands that terminate in a modified nucleotide. Analysis of fragments is similar for both methods and involves gel electrophoresis and autoradiography or fluorescence. The enzymatic method has largely replaced the chemical method as the technique of choice, although there are some situations where chemical sequencing can provide data more easily than the enzymatic method.

**DNA, recombinant:** DNA that has been cut and spliced back together in a new sequence. The DNA may be from one organism or from more than one organism

**DNA, repetitive:** Fragments of DNA that appear in multiple copies in a single individual.

**DNA:** A molecule encoding genetic information, found in the cell's nucleus. Deoxyribose Nucleic Acid. DNA is a molecule used in biological systems to store genetic information. It has a long, double stranded structure with the two strands wrapped around each other to form a right handed helix. There are about 10 nucleotide pairs per helical turn. Each strand is composed of a sugar phosphate backbone and attached bases. It is connected to a complementary strand by hydrogen bonding (non-covalent) between paired bases, adenine (A) with thymine (T) and guanine (G) with cytosine (C). DNA is an abbreviation of DeoxyriboNucleic Acid. : It is the genetic 'blueprint' of all life on this planet. : DNA typing is often applied to Human Identification which has also been called DNA Fingerprinting. DNA Fingerprinting has been attributed to Alex Jeffries, who with his colleagues in 1985, discovered a sequence of nucleotides, subsequently termed short tandem repeats (STR) when comparing the DNA sequence of the human á-globin chromosomes isolated from different individuals.

**DNA:** Deoxyribonucleic acid, the molecule that stores genetic information. Composed of two complementary strands. See base pairs.

**Dominant gene:** A gene whose phenotype is when it is present in a single copy.

**Dominant(-acting) oncogene:** A gene that stimulates cell proliferation and contributes to oncogenesis when present in a single copy. (See Oncogene.)

**Dominant:** An allele is said to be dominant if it expresses its phenotype even in the presence of a recessive allele. See Allele, Phenotype, Recessive.

**Dormancy:** A period in which a plant does not grow, awaiting necessary environmental conditions such as temperature, moisture, nutrient availability.

**Double helix:** Describes the coiling of the antiparallel strands of the DNA molecule, resembling a spiral staircase in which the paired bases form the steps and the sugar-phosphate backbones form the rails.

**Double-stranded complementary DNA (dscDNA):** A duplex DNA molecule copied from a cDNA template.

**Downstream:** The region extending in a 3' direction from a gene.

***E: coli (Escherichia coli):*** A common intestinal bacterium that is widely used in genetic engineering as a host for a cloning vector. Some strains of E. coli are important foodborne pathogens. Lab strains are of the harmless variety.

**Ecology:** The study of the interactions of organisms with their environment and with each other.

**Ecosystem:** The organisms in a plant population and the biotic and abiotic factors which impact on them. See abiotic factors; Biotic factors.

**Electrophoresis:** The technique of separating charged molecules in a matrix to which is applied an electrical field. (See Agarose gell electrophoresis, Polycrylamide gell electrophoresis.)

**Electroporation:** A method for transforrning DNA, especially useful for plant cells, in which high voltage pulses of electricity are used to open pores in cell membranes, through which foreign DNA can pass.

**Electroporation:** Transformation technique that uses electric fields to temporarily increase permeability of cells to foreign DNA.

**Emergence:** A complex whole created by simple parts, as in the brain where billions of neurons work individually, but collectively make up our consciousness and give us the ability to think, rationalize, and create.

**Emulation:** An absolutely precise simulation of something, so exact that it is equivalent to the original (for example, many computers emulate obsolete computers to run their programs). [AS] The Star Trek replicator is an example.

**Enabling science and technologies:** Areas of research relevant to a particular goal, such as nanotechnology. Also, technology that "enables" other technology to advance, such as the transistor enabled the computer chip revolution, as did photolithography.

**Encapsidation:** Process by which a virus' nucleic acid is enclosed in a capsid. See Coat protein.

**Endocrine cell:** Specialized animal cell that secretes a hormone into the blood; usually part of a gland, such as the thyroid or pituitary gland.

**Endocytosis:** Uptake of material into a cell by an invagination of the plasma membrane and its internalization in a membrane-bounded vesicle.

**Endophyte:** An organism that lives inside another.

**Endoplasmic reticulum (ER):** Labyrinthine, membrane-bounded compartment in the cytoplasm of eucaryotic cells, where lipids are synthesized and membrane-bound proteins are made.

**Endosome:** Membrane-bounded organelle in animal cells that carries materials newly ingested by endocytosis and passes many of them on to lysosomes for degradation.

**Endothelium:** Single sheet of highly flattened cells (endothelial cells) that forms the lining of all blood vessels. Regulates exchanges between the bloodstream and surrounding tissues and is usually surrounded by a basal lamina.

**Energy [J]:** Capacity for performing work or to cause heat flow. Like work itself, it is measured in Joules.

**Environmental Protection Agency (EPA):** The U.S. regulatory agency for biotechnology of microbes. The major laws under which the agency has regulatory powers are the Federal Insecticide, Fungicide, and Rodenticide Act (FIFRA); and the Toxic Substances Control Act (TSCA).

**Enzymes:** Proteins that control the various steps in all chemical reactions.

**EPROM:** Electrically Programmable Read-Only Memory — nonvolatile memory device.

**Escherichia coli:** A commensal bacterium inhabiting the human colon that is widely used in biology, both as a simple model of cell biochemical function and as a host for molecular cloning experiments.

**Ethidium bromide:** A fluorescent dye used to stain DNA and RNA. The dye fluoresces when exposed to UV light.

**Eukaryote:** An organism whose cells possess a nucleus and other membrane-bound vesicles, including all members of the protist, fungi, plant and animal kingdoms; and excluding viruses, bacteria, and blue-green algae. See Prokaryote.

**Eurisko:** A computer program developed by Professor Douglas Lenat which is able to apply heuristic rules for performing various tasks, including the invention of new heuristic rules.

**Event:** An "event" in genetic engineering is the insertion of a particular piece of foreign DNA into the chromosome of the recipient. Insertion occurs in random locations, so each event is unique. The event can affect how a gene is expressed in the organism. Once an event occurs, the transgene can be passed to the next generation as a normally inherited gene.

**Evolution:** The long-term process through which a population of organisms accumulats genetic changes that enable its members to successfully adapt to environmental conditions and to better exploit food resources.

**Exon:** A DNA sequence that is ultimately translated into protein. See DNA.

**Expectation value:** The number of different alignents with scores equivalent to or better than S that are expected to occur in a database search by chance. The lower the E value, the more significant the score.

**Express:** To translate a gene's message into a molecular product.

**Face-centered cubic:** A crystal structure found in common elemental metals also FCC. Within the cubic unit cell, atoms are located at all corner and face-centered positions.

**Fail-stop:** Describes a component or subsystem that, in the event of a failure, produces no output (e.g., of material or data) rather than producing a damaged or incorrect output.

**Fermi energy:** The energy level in a solid at which the probability of finding an electron is 1/2. For a metal, the energy corresponding to the highest filled electron state in the valence band at 0 K.

**Field-Effect Transistor (FET):** Field-Effect Transistor — semiconductor device whose insulated gate electrode controls current flow.

**FIFRA:** The Federal Insecticide, Fungicide, and Rodenticide Act. See Environmental Protection Agency.

**Fingerprint:** A set of molecular markers sufficiently diverse to identify particular individuals with reasonable certainty.

**Flanking region:** The DNA sequences extending on either side of a specific locus or gene.

**Food and Drug Administration (FDA):** The U.S. agency responsible for regulation of biotechnology food products. The major laws under which the agency has regulatory powers include the Food, Drug, and Cosmetic Act; and the Public Health Service Act.

**Fusion gene:** A hybrid gene created by joining portions of two different genes (to produce a new protein) or by joining a gene to a different promoter (to alter or regulate gene transcription).

**Gamete:** A haploid sex cell, egg or sperm, that contains a single copy of each chromosome.

**Ganglioside:** Any glycolipid having one or more sialic acid residues in its structure. Found in the plasma membrane of eucaryotic cells and especially abundant in nerve cells.

**Gel, agarose:** A substance used to separate DNA or RNA fragments by size. Used for southern and northern blots.

**Gel, polyacrylamide:** A substance used to separate proteins by size. Used for western blots.

**GEM:** A genetically engineered microorganism.

**Gene amplification:** The presence of multiple genes. Amplification is one mechanism through which proto-oncogenes are activated in malignant cells.

**Gene cloning:** The process of synthesizing multiple copies of a particular DNA sequence using a bacteria cell or another organism as a host. See DNA, Host.

**Gene expression:** The process of producing a protein from its DNA- and mRNA-coding sequences.

**Gene flow:** The exchange of genes between different but (usually) related populations.

**Gene frequency:** The percentage of a given allele in a population of organisms. See Allele.

**Gene gun:** Transformation technique that uses accelerated particles coated with DNA to introduce foreign DNA into recipient plant.

**Gene insertion:** The addition of one or more copies of a normal gene into a defective chromosome.

**Gene linkage:** The hereditary association of genes located on the same chromosome.

**Gene modification:** The chemical repair of a gene's defective DNA sequence. See DNA.

**Gene pool:** The totality of all alleles of all genes of all individuals in a particular population.

**Gene therapy:** The introduction of new genes into individuals to cure diseases or genetic abnormalities.

**Gene translocation:** The movement of a gene fragment from one chromosomal location to another, which often alters or abolishes expression.

**Gene:** The basic unit of inheritance. A segment of DNA that codes for a particular protein. A locus on a chromosome that encodes a specific protein or several related proteins. It is considered the functional unit of heredity. (See Dominant gene, Fusion gene, Gene amplification, Gene expression, Gene flow, Gene pool, Gene splicing, Gene translocation, Recessive gene, Regulatory gene.)

**Genetic assimilation:** Eventual extinction of a natural species as massive pollen flow occurs from another related species and the older crop becomes more like the new crop. See Gene flow.

**Genetic code:** The three-letter code that translates nucleic acid sequence into protein sequence. The relationships between the nucleotide base-pair triplets of a messenger RNA molecule and the 20 amino acids that are the building blocks of proteins. See Base pair, Nucleic acid, Nucleotide. The genetic information in DNA is encoded with four different nucleotide bases: A, C, G, and T (see base pair). A set of three consecutive nucleotide bases constitutes a codon. A codon specifies a particular amino acid that is added during synthesis of a protein.

**Genetic disease:** A disease that has its origin in changes to the genetic material, DNA. Usually refers to diseases that are inherited in a Mendelian fashion, although noninherited forms of cancer also result from DNA mutation.

**Genetic drift:** Random variation in gene frequency from one generation to another.

**Genetic engineering:** Genetic engineering is the process of manually adding new DNA on a molecular level with the goal of adding one or more new traits that are not already found in that organism. For the purposes of this web site, genetic engineering includes recombinant DNA and gene splicing technologies.

**Genetic linkage map:** A linear map of the relative positions of genes along a chromosome. Distances are established by linkage analysis, which determines the frequency at which two gene loci become separated during chromosomal recombination. (See Mapping.)

**Genetic marker:** A gene or group of genes used to "mark" or track the action of microbes.

**Genetic modification:** The selective and purposeful alteration of genes by humans. Genetic modification includes methods of incorporating new genes or traits into an organism through genetic engineering, mutagenesis and traditional breeding methods.

**Genome:** The genetic complement contained in the chromosomes of a given organism, usually the haploid chromosome state. The full chromosome set containing all the genes of a particular individual.

**Genomic library:** A library composed of fragments of genomic DNA. (See Library.)

**Genomics::** The study of the structure and function of genomes. Genomics usually involves high speed sequencing of the DNA and computer searches for sequences that code for genes.

**Genotype:** The sum total of the genetic information of an organism including the linkage relationships between genes. The genotype, modified by environment, determines the phenotype.

**Genus:** A category including closely related species. Interbreeding between organisms within the same category can occur.

**Germ cell (germ line) gene therapy:** The repair or replacement of a defective gene within the gamete-forming tissues, which produces a heritable change in an organism's genetic constitution.

**GMO:** Genetically modified organism. An organism that has incorporated a functional foreign gene through recombinant DNA technology. The novel gene exists in all of its cells and is passed through to progeny. Same as transgenic.

**Green revolution:** Advances in genetics, petrochemicals, and machinery that culminated in a dramatic increase in crop productivity during the third quarter of the 20th century.

**Growth curve:** See Growth phase.

**Growth factor:** A serum protein that stimulates cell division when it binds to its cell-surface receptor.

**Growth phase (curve):** The characteristic periods in the growth of a bacterial culture, as indicated by the shape of a graph of viable cell number versus time. (See Death phase, Lag phase, Logarithmic phase, Stationary phase.)

**GUI:** Graphical User Interface — hardware, software, and firmware that produces the display on modern personal computers.

**Guy Fawkes Scenario:** If nanotechnology becomes widely available, it might become trivial for anyone to committ acts of terrorism (such as making nanomachines build a large amount of explosives under government buildings a la Guy Fawkes). This would either force strict control over nanotechnology (hard) or a decentralized mode of organization.

**Hair cell:** Specialized sensory epithelial cell in the ear with bundles of giant microvilli (stereocilia) protruding from its apical surface. Sound vibrations tilt the stereocilia, evoking an electrical change in the hair cell, which thus acts as a sound detector.

**Hall effect:** The phenomenon whereby a force is applied to a moving electron or hole by a magnetic field that is applied perpendicular to the direction of motion. The force direction is perpendicular to both the magnetic field and the particle motion directions.

**Haploid cell:** A cell containing only one set, or half the usual (diploid) number, of chromosomes.

**Heme:** An iron complex. Cyclic organic molecule containing an iron atom that carries oxygen in hemoglobin and carries an electron in cytochromes.

**Hemophilia:** An X-linked recessive genetic disease, caused by a mutation in the gene for clotting factor VIII (hemophilia A) or clotting factor IX (hemophilia B), which leads to abnormal blood clotting.

**Herbicide resistance:** Herbicide resistance is when plants become resistant to the negative effects of a particular herbicide formulation or are genetically engineered in order to have resistance to a particular herbicide. Herbicide-resistant crops can be sprayed with a particular herbicide, killing weeds growing in and around the crop plants, without being damaged.

**Herbicide:** Any substance that is toxic to plants; usually used to kill specific unwanted plants.

**Heterochromatin:** Dark-stained regions of chromosomes thought to be for the most part genetically inactive.

**Heteroduplex:** A double-stranded DNA molecule or DNA-RNA hybrid, where each strand is of a different origin.

**Heterogeneous nuclear RNA (hnRNA):** The name originally given to large RNA molecules found in the nucleus, which are now known to be unedited mRNA transcripts, or pre-mRNAs. (See RNA.)

**Homologous chromosomes:** Chromosomes that have the same linear arrangement of genes—a pair of matching chromosomes in a diploid organism. See Chromosomes.

**Homologous recombination:** The exchange of DNA fragments between twc DNA molecules or chromatids of paired chromosomes (during crossing over) at the site of identical nucleotide sequences.

**Homozygote:** An organism whose genotype is characterized by two identical alleles of a gene. See Allele, Genotype.

**Human Genome Project:** A project coordinated by the National Institutes of Health (NIH) and the Department of Energy (DOE) to determine the entire nucleotide sequence of the human chromosomes. (See NIH.)

**Human growth hormone (HGH, somatotrophin):** A protein produced in the pituitary gland that stimulates the liver to produce somatomedins, which stimulate growth of bone and muscle.

**Hybrid:** The offspring of two parents differing in at least one genetic characteristic (trait). Also, a heteroduplex DNA or DNA-RNA molecule.

**Hybridization, nucleic acid:** Complementary strands of DNA or RNA will spontaneously match up with each other and bond together under the right conditions. This is called nucleic acid hybridization and is used when probing southern and northern blots.

**Hybridization:** The hydrogen bonding of complementary DNA and/or RNA sequences to form a duplex molecule. (See Northern hybridization, Southern hybridization.)

**Hybridoma:** A hybrid cell, composed of a B Iymphocyte fused to a tumor cell, which grows indefinitely in tissue culture and is selected for the secretion of a specific antibody of interest.

**Hydrogen bond:** A relatively weak bond formed between a hydrogen atom (which is covalently bound to a nitrogen or oxygen atom) and a nitrogen or oxygen with an unshared electron pair.

**Hydrolysis:** A reaction in which a molecule of water is added at the site of cleavage of a molecule to two products.

**Imaging Contrast Agent:** A molecule or molecular complex that increases the intensity of the signal detected by an imaging technique, including MRI and ultrasound. An MRI contrast agent, for example, might contain gadolinium attached to a targeting antibody. The antibody would bind to a specific target—a metastatic melanoma cell, for example – while the gadolinium would increase the magnetic signal detected by the MRI scanner.

**Immortalization:** Production of a cell line capable of an unlimited number of cell divisions. Can be the result of a chemical or viral transformation or of fusion with cells of a tumor line.

**Immortalizing oncogene:** A gene that upon transfection enables a primary cell to grow indefinitely in culture. (See Oncogene.)

**In situ:** Refers to performing assays or manipulations with intact tissues.

**In vivo:** Refers to biological processes that take place within a living organism or cell.

**Incomplete dominance:** A condition where a heterozygous off- spring has a phenotype that is distinctly different from, and intermediate to, the parental phenotypes. See Heterozygote, Phenotype.

**Initiation codon:** The mRNA sequence AUG, coding for methionine, which initiates translation of mRNA.

**Inositol lipid:** A membrane-anchored phospholipid that transduces hormonal signals by stimulating the release of any of several chemical messengers. (See Phospholipid.)

**insect resistance management:** A set of strategies to reduce the frequency of resistance in crop pests. These strategies are widely referred to as Insect Resistance Management.

**Insertion mutations:** Changes in the base sequence of a DNA molecule resulting from the random integration of DNA from another source. See DNA, Mutation.

**Insulator:** A substance, object or material that does not conduct electricity. Material that conducts electricity very poorly.

**Insulin:** A peptide hormone secreted by the islets of Langerhans of the pancreas that regulates the level of sugar in the blood.

**Integrated Circuit (IC):** An electronic circuit consisting of many interconnected devices on one piece of semiconductor, typically into 10 millimeters on a side. ICs are the major building blocks of today's computers. Semiconductor circuit, typically on a very small silicon chip, containing microfabricated transistors, diodes, resistors, capacitors, etc.

**Interferon:** A family of small proteins that stimulate viral resistance in cells.

**Intergenic regions:** DNA sequences located between genes that comprise a large percentage of the human genome with no known function.

**Introgression:** Backcrossing of hybrids of two plant populations to introduce new genes into a wild population.

**Intron:** A noncoding DNA sequence within a gene that is initially transcribed into messenger RNA but is later snipped out. See Coding, DNA, Messenger RNA, Transcription.

**Invasiveness:** Ability of a plant to spread beyond its introduction site and become established in new locations where it may provide a deliterious effect on organisms already existing there.

**Ion:** A charged particle.

**Isotope:** One of two or more forms of an element that have the same number of protons (atomic number) but differing numbers of neutrons (mass numbers). Radioactive isotopes are commonly used to make DNA probes and metabolic tracers.

**Joining (J) segment:** A small DNA segment that links genes to yield a functional gene encoding an immunogobulin.

**Jupiter-Brain:** A posthuman being of extremely high computational power and size. This is the archetypal concentrated intelligence. The term originated due to an idea by Keith Henson that nanomachines could be used to turn the mass of Jupiter into computers running an upgraded version of himself.

**Kanamycin:** An antibiotic of the aminoglycoside family that poisons translation by binding to the ribosomes.

**Karyotype:** All of the chromosomes in a cell or an individual organism, visible through a microsope during cell division.

**Kb (kilobase):** The number of base pairs (in thousands) that denote the size of a nucleic acid fragment.

**Kilohertz (kHz):** One thousand cycles per second.

**Kilojoule:** Standard unit of energy equal to 1000 joules, or 0.24 kilocalories.

**Lag phase:** The initial growth phase, during which cell number remains relatively constant prior to rapid growth. See growth phase.

**Lane:** On a gel or a blot, a lane belongs to one sample. It contains fragments of different sizes that were separated by gel electrophoresis. See band, blot.

**Lawn:** A uniform and uninterrupted laver of bacterial growth, in which individual colonies cannot be observed.

**Lectin:** Protein that binds tightly to a specific sugar. Abundant lectins derived from plant seeds are often used as affinity reagents to purify glycoproteins or to detect them on the surface of cells.

**Legume:** A member of the pea family that possesses root nodules containing nitrogen-fixing bacteria.

**Library, DNA:** A large set of clones of DNA fragments from a particular organism. DNA libraries are usually maintained in *E. coli* or in bacteriophages.

**Library:** A collection of cells, usually bacteria or yeast, that have been transformed with recombinant vectors carrying DNA inserts from a single species. (See cDNA library, Expression library, Genomic library.)

**Life (lifetime):** The length of time the sensor can be used before its performance changes.

**Life Science Applications of Deep Red T2-MP EviTag Quantum Dots:** The new T2-MP EviTags are available in the deep red wavelengths used in life science research – Macoun Red (660nm) and Jonamac Red (680nm) – with carboxyl, amine, biotin and non-functionalized surfaces. Having a biologically compatible quantum dot label like the T2-MP EviTags allows further exploration and development of in vivo assays, following cells inside living organism to get a new understanding of life. In particular, it offers the possibility to create new classes of assays for use where current testing materials could produce cytotoxicity or cell death. Potential applications include live cell and even whole animal imaging, blood cancer assays, and numerous other applications.

**Ligand:** In protein chemistry, a small molecule that is (or can be) bound by a larger molecule is termed a ligand. In organometallic chemistry, a moiety bonded to a central metal atom is also termed a ligand; the latter definition is more common in general chemistry.

**Ligase (DNA ligase):** An enzyme that catalyzes a condensation reaction that links two DNA molecules via the formation of a phosphodiester bond between the 3' hydroxyl and 5' phosphate of adjacent nucleotides.

**Ligate:** The process of joining two or more DNA fragments.

**Ligation:** Joining two fragments of DNA end to end. See sticky ends.

**Light Emitting Diodes ( LEDs):** Light Emitting Diodes work on a completely different concept. Traditionally LEDs are created from two semiconductors. By running current in one direction across the semiconductor the LED emits light of a particular frequency (hence a particular color) depending on the physical characteristics of the semiconductor used. The semiconductor is covered with a piece of plastic that focuses the light and increases the brightness. These semiconductors are very durable, there is no filament, they donít require much power, theyíre brighter and they last a long time. By densely packing red, blue and green LEDs next to each other on a substrate one can create a display.

**Lineage:** A chart that traces the flow of genetic information from generation to generation.

**Linkage:** The frequency of coinheritance of a pair of genes and/or genetic markers, which provides a measure of their physical proximity to one another on a chromosome. Measures the physical distance between two genes. Genes that are close together are unlikely to segregate in a sexual cross. Distant genes segregate independently are are then said to be unlinked.

**Linked genes/markers:** Genes and/or markers that are so closely associated on the chromosome that they are coinherited in 80% or more of cases.

**Linker:** A short, double-stranded oligonucleotide containing a restriction endonuclease recognition site, which is ligated to the ends of a DNA fragment.

**Liposome:** A type of nanoparticle made of lipids, or fat molecules, surrounding a water core. Liposomes, several of which are widely used to treat infectious diseases and cancer, were the first type of nanoparticle to be used to create therapeutic agents with novel characteristics. Membrane-bound vesicles constructed in the laboratory to transport biological molecules.

**Liquid Crystal Display:** LCD is the predominant technology used in flat panel displays. The principle that makes the display work is this: A crystalís alignment can be altered with an electric current. If the crystal is lined up one way ñ it will allow the light waves to pass through a polarized filter, but if the electric current alters the crystalís alignment, it will guide light so that the polarized filter blocks the light. By densely packing red, blue and green light emitting crystals next to each other on a sheet (ìcalled a substrateî), one can create a full color display. The great thing about LCD is that the crystals can be packed together closely, allowing for a higher-resolution, finer-detail display. The con is that LCDs are somewhat fragile, require a lot of power and are relatively less bright. Liquid Crystal Display — display device employing light source and electrically alterable optically active thin film.

**Locus (plural = loci):** A specific location or site on a chromosome.

**Logarithmic phase (log or exponential growth phase):** The steepest slope of the growth curve—the phase of vigorous growth during which cell number doubles every 20-30 minutes. (See Growth phase.)

**Lysogen:** A bacterial cell whose chromosome contains integrated viral DNA.

**Lysogenic:** A type or phase of the virus life cycle during which the virus integrates into the host chromosome of the infected cell, often remaining essentially dormant for some period of time. See Lysogen.

**Lytic:** A phase of the virus life cycle during which the virus replicates within the host cell, releasing a new generation of viruses when the infected cell lyses.

**Machine-phase chemistry:** The chemistry of systems in which all potentially reactive moieties follow controlled trajectories (e.g., guided by molecular machines working in vacuum).

**Macroscale:** Larger than nanoscale; often implies a design that humans can directly interact with; too large to be built by a single assembler (one cubic micron of diamond contains 176 billion atoms).

**Magnetization (M):** The total magnetic moment per unit volume of material. Also, a measure of the contribution to the magnetic flux by some material within an H field. The magnitude of M is proportional to the applied field as: $M = \div_m \times H$, with $\div_m$ the magnetic susceptibility.

**Magnetostrictive material:** A material that changes dimension in the presence of a magnetic field or generates a magnetic field when mechanically deformed.

**Malignant:** Having the properties of cancerous growth.

**Manufacturing:** The ability to make products, in this case ranging from clothing, to electronics, to medical devices, to books, to building materials, and much more.

**Mapping:** Determining the physical location of a gene or genetic marker on a chromosome. **Measurand:** A physical quantity, condition, or property that is to be measured.

**Meat Machine:** AKA Cabinet Beast. A box containing assemblers and raw material, within which is formed meat [or whatever else it was programmed to make].

**Mechanical:** Pertaining to the positions and motions of atoms, as defined by the positions of their nuclei; see electronic. A purely mechanical device can be described in terms of atomic positions and motions without reference to electronic properties, save through their effect on the potential energy function.

**Mechanochemistry:** Chemistry accomplished by mechanical systems directly controlling the reactant molecules; the formation or breaking of chemical bonds under direct mechanical control. In this volume, the chemistry of processes in which mechanical systems operating with atomic-scale precision either guide, drive, or are driven by chemical transformations. In general usage, the chemistry of processes in which energy is converted from mechanical to chemical form, or vice versa.

**Mechanosynthesis:** Chemical synthesis controlled by mechanical systems operating with atomic-scale precision, enabling direct positional selection of reaction sites; synthetic applications of mechanochemistry. Suitable mechanical systems include AFM mechanisms, molecular manipulators, and molecular mill systems. Processes that fall outside the intended scope of this definition include reactions guided by the incorporation of reactive moieties into a shared covalent framework (i.e., conventional intramolecular reactions), or by the binding of reagents to enzymes or enzymelike catalysts. Molecular tools with chemically specific tip structures can be used, sequentially, to modify a work piece and build a wide range of molecular structures.

**Mechatronics:** The study of the melding of AI and electromechanical machines to make machines that are greater than the sum of their parts. The synergistic combination of precision mechanical engineering with electronic control.

**Megabase cloning:** The cloning of very large DNA fragments. (See Cloning.)

**Megabyte (MB):** $2^{20}$ (= 1,048,576, or about one million) bytes of information.

**Megahertz (Mhz):** One million cycles per second

**Meiosis:** The reduction division process by which haploid gametes and spores are formed, consisting of a single duplication of the genetic material followed by two mitotic divisions.

**Meme:** An idea that replicates through a society as it is propagated through person-to-person interaction, both direct and indirect. Memetics is a field of study that focuses on memes' role in the evolution of a culture. Self-reproducing idea or other information pattern which is propagated in ways similar to that of a gene. [Richard Dawkins, 1976]

**Messenger RNA (mRNA):** The class of RNA molecules that copies the genetic information from DNA, in the nucleus, and carries it to ribosomes, in the cytoplasm, where it is translated into protein. (See RNA.)

**Metallothionein:** A protective protein that binds heavy metals, such as cadmium and lead.

**Micellar nanocontainers:** Block copolymer micelles are water- soluble biocompatible nanocontainers with great potential for delivering hydrophobic drugs. An understanding of their cellular distribution is essential to achieving selective delivery of drugs at the subcellular level. R. Savic, L. Luo, A. Eisenberg, D. Maysinger, Micellar nanocontainers distribute to defined cytoplasmic organelles Science 300 (5619): 615- 618, Apr. 25, 2003

**Micelles:** Micelles are small, spherical structures composed of molecules that attract one another to reduce surface tension. The head of the molecule is hydrophilic, meaning it likes water, while the interior portion is hydrophobic, meaning it avoids water.

**Microbial mats (biofilms):** Layered groups or communities of microbial populations.

**Microelectronics:** Narrower terms: MEMS, nanoelectronics, optoelectronics, SED Single Electron Devices. Related terms: molecular electronics, semiconductors

**Microencapsulation:** Individually encapsulated small particles.

**Microinjection:** A means to introduce a solution of DNA, protein, or other soluble material into a cell using a fine microcapillary pipet.

**Microspectrophotometry:** Analytical technique for studying substances present at enzyme concentrations in single cells, in situ, by measuring light absorption. Light from a tungsten strip lamp or xenon arc dispersed by a grating monochromator illuminates the optical system of a microscope. The absorbance of light is measured (in nanometers) by comparing the difference between the image of the sample and a reference image. MeSH 1990

**Microspheres:** Microspheres are generally defined as small spheres made of any material and sized from about 0.5 $\mu$m to 100 $\mu$m. Similar, but smaller spheres sized 10 to 500 nm are called *nanospheres.* Ideally, microspheres are completely spherical and homogeneous in size, although less perfect particles are often termed microspheres as well. Depending on the preparation method and material used, microspheres show a typical size distribution which often deviates from the mono-sized ideal.

**Microstructure:** In material engineering the structural features of a material such as grain boundaries, grain size and structure, subject to observation under a microscope, selective etching etc. In MEMS microstructure unfortunately is also used to designate a micromachined feature. The last decade has seen rapid developments in the fabrication, characterization and conceptual understanding of synthetic microstructures in many different material systems including silicon, III-V and II-VI semiconductors, metals, ceramics and organics. The objective of this journal [Superlattices and Microstructures] is to provide a common interdisciplinary platform for the publication of the latest research results on all such "nanostructures" with dimensions in the range of 1-100 nm; the unifying theme here being the dimensions of these artificial structures rather than the material system in which they are fabricated.

**Microsystem:** A microscale machine that can sense information from the environment and act accordingly. Outside the U.S., it can also refer to microelectromechanical systems (MEMS). [smalltimes glossary, 2002]

**MicroTas, micro Total Analysis Systems, uTAS:** Although initial research dates back to the early 1970's, the field of micro- TAS formally started in 1990, when Manz et al described the possibility of creating microsystems that would take care of many or all the traditional analytical steps involved in a biochemical analysis (sample introduction, handling, extraction, purification, concentration, filtration, analysis, detection) .... Micro- TAS offer many advantages over traditional analysis systems. Low power consumption and small reaction volumes, faster analysis, ultrasensitive detection, and minimal human intervention are key parameters in the development of micro- TAS. Most biochemical reactions take place in liquid environments. Hence, the development of MicroTAS is intrinsically linked to the design of liquid handling micro- devices. [Biomedical Applications Group (GAB) Centro Nacional de Microelectronica (CNM- IMB) Bellaterra, Spain, 2000]

**Microtubule:** A hollow cylindrical tube used for the transport of materials inside cells.

**Miller indices:** A set of 3 integers (4 for hexagonal) that designate crystallographic planes, as determined from reciprocals of fractional axial intercepts.

**Millimeter:** One thousandth of a meter, or about 1/26 of an inch.

**Mitosis:** The replication of a cell to form two daughter cells with identical sets of chromosomes.

**Molecular - Nanotechnology:** The technology of precisely-constructed molecular-scale machines; from nanometer: a billionth of a meter. For a good introduction, see What is Nanotechnology? or Ralph C. Merkle's nanotechnology page.

**Molecular assembler:** A general-purpose device for molecular manufacturing, able to guide chemical reactions by positioning individual molecules to atomic accuracy (e.g. mechanosynthesis) and to construct a wide range of useful and stable molecular structures according to precise specifications. Also known as an assembler, a molecular assembler is a molecular machine that can build a molecular structure from its component building blocks. [ZY]

**Molecular Beam Epitaxy:** [MBE] Process used to make compound (multi-layer) semiconductors. Consists of depositing alternating layers of materials, layer by layer, one type after another (such as the semiconductors gallium arsenide and aluminum gallium arsenide).

**Molecular biology:** The study of the biochemical and molecular interactions within living cells.

**Molecular cloning::** The biological amplification of a specific DNA sequence through mitotic division of a host cell into which it has been transformed or transfected. (See Cloning.)

**Molecular genetics:** The study of the flow and regulation of genetic information between DNA, RNA, and protein molecules.

**Molecular markers:** Genetic traits which are detectable on gels or blots and can be used to construct genetic maps. Molecular markers usually have no known function. RFLPs are molecular markers.

**Molecular mechanics models:** Many of the properties of molecular systems are determined by the molecular potential energy function. Molecular mechanics models approximate this function as a sum of 2-atom, 3-atom, and 4-atom terms, each determined by the geometries and bonds of the component atoms. The 2-atom and 3-atom terms describing bonded interactions roughly correspond to linear springs.

**Molecular medicine:** A variety of pharmaceutical techniques and gene therapies that address specific molecular diseases or molecular defects in biological systems. A variety of pharmaceutical techniques and therapies in use today. Studying molecules as they relate to health and disease, and manipulating those molecules to improve the diagnosis, prevention, and treatment of disease.

**Molecular mill:** A mechanochemical processing system characterized by limited motions and repetitive operations without programmable flexibility.

**Molecular motors:** Protein based machines that are involved in or cause movement such as the rotary devices (flagellar motor and the F1 ATPase) or the devices whose movement is directed along cytoskeletal filaments (myosin, kinesin and dynein motor families). [MeSH, 1999]

**Molecular Nanoscience:** An emerging interdisciplinary field that combines the study of molecular/ biomolecular systems with the science and technology of nanoscale structures and systems. The potential applications for this research are very broad and include such possibilities as 1) the use of biomolecules and cellular systems to self- assemble nanoelectronic circuitry and other nanoscale structures and 2) the use of lamellar host frameworks containing nano pores that can be tailored to include guest molecules for separation of chemicals for pharmaceutical and other applications. Molecular Nanoscience Alliance for Interdisciplinary Studies and Activities, Univ. of Minnesota, US,

**Molecular Wire:** A molecular wire—the simplest electronic component—is a quasi-one-dimensional molecule that can transport charge carriers (electrons or holes) between its ends. [Michael D Ward]

**Molecularly imprinted polymers MIPs:** A new class of materials that have artificially created receptor structures. Since their discovery in 1972, MIPs have attracted considerable interest from scientists and 3engineers involved with the development of chromatographic absorbents, membranes, sensors and enzyme and receptor mimics.

**Monoclonal antibodies:** Immunoglobulin molecules of single- epitope specificity that are secreted by a clone of B cells.

**Monoculture:** The agricultural practice of cultivating crops consisting of genetically similar organisms.

**Multi-locus probe:** A probe that hybridizes to a number of different sites in the genome of an organism. (See Probe.)

**Mutagen:** Any agent or process that can cause mutations. See Mutation.

**Mutation:** An alteration in DNA structure or sequence of a gene. A spontaneous or induced genetic change in the DNA of an organism. Most mutations are undesirable, but a few are useful such as dwarfing genes in cereal crops that allow the plants to stand better.

**Mycorrhizae:** Fungi that form symbiotic relationships with roots of more developed plants.

**Nanobiology:** Many fundamental biological functions are carried out by molecular machineries that have the sizes of 1-100 nm. You find many examples in molecular biology and cell biology: single enzymes, transcription complex, ribosome, transport complex, nuclear pore, and so on. To understand the functions of these machineries, one has to describe their movements, changes in their shapes, and their localization. This means the mechanistic study is equivalent to dynamic morphology at this level of size, making a new field that merges mechanistic biology and morphology. The emergence of nanobiology depended on the invention of nano-technology: scanning probe microscopy, modern optical techniques, and micro- manipulating techniques. This concept of nanobiology was first proposed in a group study named "Biological Nano-Mechanisms", which was supported by Japanese Agency of Science and Technology (1992-1997). [National Institute of Genetics, Japan]

**Nanobiotechnology:** An emerging area of scientific and technological opportunity. Nanobiotechnology applies the tools and processes of nano/ microfabrication to build devices for studying biosystems. Researchers also learn from biology how to create better micro-nanoscale devices. The Nanobiotechnology Center (NBTC), a National Science Foundation, Science and Technology Center is characterized by its highly interdisciplinary nature and features a close collaboration between life scientists, physical scientists, and engineers. [NBTC, NanoBiotechnology Center, Cornell Univ. US, 2002] Use of nanotechnology[ies] in the life sciences. These may include, but are not limited to, therapeutics, medical devices/implants, biosensors, and tools for the development of drugs. Applying the tools and processes of MNT to build devices for studying biosystems, in order to learn from biology how to create better nanoscale devices. Should hasten the creation of useful micro devices that mimic living biological systems.

**Nanocentrifuge:** In medical nanorobotics, a proposed nanodevice that can spin materials at very high speed, imparting rotational accelerations of up to one trillion gravities (g's), thus permitting rapid sortation.

**Nanoscience:** Nanoscience is the science of nanoscale phenomena. This is a very complicated area, since Quantum Mechanics is often needed to describe the physical properties of atoms and molecules properly. What makes it even more complicated is that nanostructured materials can still be quite a challenge to describe Quantum Mechanically, since the equations are feindishly difficult—this is where supercomputers come in to make the job approachable. The scientific discipline seeking to increase our knowledge and understanding of nanoscale phenomena, i.e. science on the scale of 0.1 nm to 100 nm.

**Nanosensor:** A chemical or physical sensor constructed using nanoscale components, usually microscopic or submicroscopic in size.

**Nanosources**: sources that emit light from nanometre-scale volumes.

**Nanotech:** Slang for nanotechnology.

**National Institutions of Health (NIH):** A nonregulatory agency which has oversight of research activities that the agency funds.

**National Science Foundation (NSF):** A nonregulatory agency which has oversight of biotechnology research activities that the agency funds.

**Natural selection:** The differential survival and reproduction of organisms with genetic characteristics that enable them to better utilize environmental resources.

**Neophile:** One who welcomes the future and who enjoys change and evolution.

**Nick translation:** A procedure for making a DNA probe in which a DNA fragment is treated with DNase to produce single-stranded nicks, followed by incorporation of radioactive nucleotides from the nicked sites by DNA polymerase I.

**Nicked circle (relaxed circle):** During extraction of plasmid DNA from the bacterial cell, one strand of the DNA becomes nicked. This relaxes the torsional strain needed to maintain supercoiling, producing the familiar form of plasmid. (See Plasmid.)

**Nitrocellulose:** A membrane used to immobilize DNA, RNA, or protein, which can then be probed with a labeled sequence or antibody.

**Nitrogen fixation:** The conversion of atmospheric nitrogen to biologically usable nitrates.

**Nitrogenous bases:** The purines (adenine and guanine) and pyrimidines (thymine, cytosine, and uracil) that comprise DNA and RNA molecules.

**Nodule:** The enlargement or swelling on roots of nitrogen-fixing plants. The nodules contain symbiotic nitrogen-fixing bacteria. See Nitrogen fixation.

**Nontarget organism:** An organism which is affected by an interaction for which it was not the intended recipient.

**Northern hybridization:** (Northern blotting). A procedure in which RNA fragments are transferred from an agarose gel to a nitrocellulose filter, where the RNA is then hybridized to a radioactive probe. (See Hybridization.)

**Novel foods:** Novel foods are products that have never been used as a food; foods which result from a process that has not previously been used for food; or, foods that have been modified by genetic manipulation. This last category of foods have been described as genetically modified foods (often referred to as GM foods, genetically engineered foods or biotechnology-derived foods). *Definition from Health Canada Web site*

**Nuclease:** Any enzyme that cuts nucleic acids. See restriction enzyme. A class of enzymes that degrades DNA and/or RNA molecules by cleaving the phosphodiester bonds that link adjacent nucleotides. In deoxyribonuclease (DNase), the substrate is DNA. In endonuclease, it cleaves at internal sites in the substrate molecule. Exonuclease progressively cleaves from the end of the substrate molecule. In ribonuclease (RNase), the substrate is RNA. In the S1 nuclease, the substrate is single-stranded DNA or RNA.

**Nucleic acids:** The two nucleic acids, deoxyribonucleic acid (DNA) and ribonucleic acid (RNA), are made up of long chains of molecules called nucleotides. See DNA, RNA, Nucleotides.

**Nuclein:** The term used by Friedrich Miescher to describe the nuclear material he discovered in 1869, which today is known as DNA.

**Nucleoside analog:** A synthetic molecule that resembles a naturally occuring nucleoside, but that lacks a bond site needed to link it to an adjacent nucleotide. (See Nucleoside.)

**Nucleotide:** A building block of DNA and RNA, consisting of a nitrogenous base, a five-carbon sugar, and a phosphate group. Together, the nucleotides form codons, which when strung together form genes, which in turn link to form chromosomes.

**Nucleus:** The membrane-bound region of a eukaryotic cell that contains the chromosomes. A cellular organelle in plants and animals that contains the chromosomes which in turn are composed of DNA plus protein.

**Occupational Safety and Health Administration (OSHA):** One of the U.S. agencies responsible for regulation of biotechnology. The major law under which the agency has regulatory powers is the Occupational Safety and Health Act.

**Oligonucleotide:** A DNA polymer composed of only a few nucleotides. (See Nucleotide.)

**Oncogene:** A gene that contributes to cancer formation when mutated or inappropriately expressed. (See Cellular oncogene, Dominant oncogene, Immortalizing oncogene, Recessive oncogene.)

**Oncogenesis:** The progression of cytological, genetic, and cellular changes that culminate in a malignant tumor.

**Open pollination:** Pollination by wind, insects, or other natural mechanisms.

**Open reading frame:** A long DNA sequence that is uninterrupted by a stop codon and encodes part or all of a protein. (See Reading frame.)

**Operator:** A prokaryotic regulatory element that interacts with a repressor to control the transcription of adjacent structural genes.

**Orbital:** In the approximation that each electron in a molecule has a distinct, independent wave function, the spatial distribution of an electron wave function corresponds to a molecular orbital. These, in turn, can be approximated as sums of contributions from the orbitals characteristic of the isolated atoms. An electron added to a molecule—or, similarly, one excited to a higher-energy state within a molecule—would occupy a state with a different wave function from the rest; an unoccupied state of this kind corresponds to an unoccupied molecular orbital. Orbital-symmetry effects on reaction rates arise when a reaction requires overlap between two lobes of the orbitals on each of two reagents: if the algebraic signs of the wave functions in the facing lobes do not match, bond formation between those orbitals is prohibited.

**Organelle:** A cell structure that carries out a specialized function in the life of a cell.

**Origin of replication:** The nucleotide sequence at which DNA synthesis is initiated.

**Overlapping reading frames:** Start codons in different reading frames generate different polypeptides from the same DNA sequence.

**Ovum:** A female gamete.

**Paleontology:** The study of the fossil record of past geological periods and of the phylogenetic relationships between ancient and contemporary plant and animal species.

**Palindromic sequence:** A DNA locus whose 5'-to-3' sequence is identical on each DNA strand. The sequence is the same when one strand is read left to right and the other strand is read right to left. Recognition sites of many restriction enzymes are palindromic. See DNA.

**pAMP:** Ampicillin-resistant plasmid developed for this laboratory course. (See Plasmid.)

**Parasitism:** The closee association of two or more dissimilar organisms where the association is harmful to at least one. See Commensalism, Parasitism, Symbiosis.

**Particle bombardment:** Using metal particles to blast DNA into cells for the purpose of genetic engineering. A gene gun is used to propel the metal particles.

**Pathogen:** Organism which can cause disease in another organism.

**pBR322:** A derivation of ColE1, one of the first plasmid vectors widely used. (See Plasmid.)

**PCR:** Polymerase chain reaction. An amazingly sensitive technique that allows the specific amplification of extremely small amounts of particular DNA fragments using DNA polymerase and specific primers.

**Pedigree:** A diagram mapping the genetic history of a particular family.

**Percent Accepted Mutation:** A unit introduced by Dayhoff et al. to quantify the amount of evolutionary change in a protein sequence. 1.0 PAM unit, is the amount of evolution which will change, on average, 1% of amino acids in a protein sequence. A PAM(x) substitution matrix is a look-up table in which scores for each amino acid substitution have been calculated based on the frequency of that substitution in closely related proteins that have experienced a certain amount (x) of evolutionary divergence.

**Persistence:** Ability of an organism to remain in a particular setting for a period of time after it is introduced.

**Pesticide:** A substance that kills harmful organisms (for example, an insecticide or fungicide).

**Phenotype:** The observable characteristics of an organism, the expression of gene alleles (genotype) as an observable physical or biochemical trait. See Genotype.

**Pheromone:** A hormone-like substance that is secreted into the environment.

**Phosphatase:** An enzyme that hydrolyzes esters of phosphoric acid, removing a phosphate group.

**Phosphodiester bond:** A bond in which a phosphate group joins adjacent carbons through ester linkages. A condensation reaction between adjacent nucleotides results in a phosphodiester bond between 3' and 5' carbons in DNA and RNA.

**Phospholipid:** A class of lipid molecules in which a phosphate group is linked to glycerol and two fatty acyl groups. A chief component of biological membranes. (See Inositol phospholipid.)

**Physical map:** A map showing physical locations on a DNA molecule, such as restriction sites, and sequence-tagged sites.

**Piezoelectric material:** A ferroelectric material in which an electrical potential difference is created due to mechanical deformation, or conversely, in which the application of a voltage causes dimensional changes in the material.

**Piezoelectric:** A piezoelectric material expands or contracts when a voltage is applied across it. Piezoelectric tubes are often used to move the tip in an AFM or STM. If a piezoelectric crystal is compressed, a voltage appears across it. This effect is sometimes used to measure displacements. A commonly used piezoelectric is PZT $(Pb,Zr)TiO_3$.

**Plaque:** A clear spot on a lawn of bacteria or cultured cells where cells have been Iysed by viral infection.

**Plasmid (p):** A circular DNA molecule, capable of autonomous replication, which typically carries one or more genes encoding antibiotic resistance proteins. Plasmids can transfer genes between bacteria and are important tools of transformation for genetic engineers. (See Nicked circle, pAMP, Relaxed plasmid, Stringent plasmid, Supercoiled plasmid.)

**Pleiotrophy:** The effect of a particular gene on several different traits.

**Point mutation:** A change in a single base pair of a DNA sequence in a gene. (See Mutation.)

**Poly(A) polymerase:** Catalyzes the addition of adenine residues to the 3' end of pre-mRNAs to form the poly(A) tail. (See Polymerase.)

**Polyacrylamide gel electrophoresis:** Electrophoresis through a matrix composed of a synthetic polymer, used to separate proteins, small DNA, or RNA molecules of up to 1000 nucleotides. Used in DNA sequencing. (See Electrophoresis.)

**Polyclonal antibodies:** A mixture of immunoglobulin molecules secreted against a specific antigen, each recognizing a different epitope.

**Polygenic:** Controlled by or associated with more than one gene.

**Polylinker:** A short DNA sequence containing several restriction enzyme recognition sites that is contained in cloning vectors.

**Polymer:** A molecule composed of repeated subunits.

**Polymerase (DNA):** Synthesizes a double-stranded DNA molecule using a primer and DNA as a template. (See Poly(A) polymerase, Polymerase chain reaction, RNA polymerase, Taq polymerase.)

**Polymerase chain reaction (PCR):** A procedure that enzymatically amplifies a DNA polymerase. (See Polymerase.)

**Polymerase:** Any enzyme that adds subunits to chains of macromolecules. Example is DNA polymerase.

**Polymorphisms:** Variant forms of a particular gene that occur simultaneously in a population.

**Polynucleotide:** A DNA polymer composed of multiple nucleotides. (See Nucleotide.)

**Polypeptide (protein):** A polymer composed of multiple amino acid units linked by peptide bonds.

**Polyploid:** A multiple of the haploid chromosome number that results from chromosome replication without nuclear division.

**Polysaccharide:** A polymer composed of multiple units of monosaccharide (simple sugar).

**Polyvalent vaccine:** A recombinant organism into which has been cloned antigenic determinants from a number of different disease-causing organisms. (See Vaccine.)

**Population:** A local group of organisms belonging to the same species and capable of interbreeding.

**Primary cell:** A cell or cell line taken directly from a living organism, which is not immortalized.

**Primer:** A short DNA or RNA fragment annealed to single-stranded DNA, from which DNA polymerase extends a new DNA strand to produce a duplex molecule.

**Probe:** A single-stranded DNA that has been radioactively labeled and is used to identify complementary sequences in genes or DNA fragments of interest. (See Multilocus probe.) A fragment of DNA, usually labeled with radioactive $^{32}$P, that detects homologous sequences on southern blots.

**Prokaryote:** A bacterial cell lacking a true nucleus; its DNA is usually in one long strand. See Eukaryote.

**Promoter:** A region of DNA extending 150-300 bp upstream from the transcription start site that contains binding sites for RNA polymerase and a number of proteins that regulate the rate of transcription of the adjacent gene. (See Constitutive promoter.)

**Pronucleus:** Either of the two haploid gamete nuclei just prior to their fusion in the fertilized ovum.

**Protease:** An enzyme that cleaves peptide bonds that link amino acids in protein molecules.

**Protein:** Molecules composed of amino acids. Proteins constitute the enzymes and many of the structural components of cells. A polymer of amino acids linked via peptide bonds and which may be composed of two or more polypeptide chains. (See Polypeptide.) Proteins do the work in cells. They can be part of structures (such as cell walls, organelles, etc), regulate reactions that take place in the cell or they can serve as enzymes, which speed-up reactions. Everything you see in an organism is either made of proteins or the result of a protein action.

**Protein kinase:** An enzyme that adds phosphate groups to a protein molecule at serine, threonine, or tyrosine residues.

**Proteinaceous infectious particle (prion):** A proposed pathogen composed only of protein with no detectable nucleic acid and which is responsible for Creutzfeldt-Jakob disease and kuru in humans and scrapie in sheep.

**Quantum Mechanics:** A largely computational physical theory explaining the behavior of quantum phenomena, which incorporates the theory of special relativity. Despite dilignet attempts, general relativity has not been sucessfully incorporated into quantum mechanics. A physical model of chemical and optical phenomena, as well as the behaviour of matter in general on a small scale. Quantum mechanics describes a system of particles in terms of a wave function defined over the configuration space of the system. Although the concept of particles having distinct locations is implicit in the potential energy function that determines the wave function (e.g., of a ground-state system), the observable dynamics of the system cannot be described in terms of the motion of such particles from point to point. In describing the energies, distributions, and behaviors of electrons in nanometer-scale structures, quantum mechanical methods are necessary. Electron wave functions help determine the potential energy surface of a molecular system, which in turn is the basis for classical descriptions of molecular motion. Nanomechanical systems can almost always be described in terms of classical mechanics, with occasional quantum mechanical corrections applied within the framework of a classical model.

**Quantum Well:** A P-N-P junction in which the "N" layer is ~10 nm (where traditional physics leaves off and quantum effects take over) and an "electron trap" is created. "If one makes a heterostructure with sufficiently thin layers, quantum interference effects begin to appear prominently in the motion of the electrons. The simplest structure in which these may be observed is a quantum well, which simply consists of a thin layer of a narrower-gap semiconductor between thicker layers of a wider-gap material."

**Reading frame:** A series of triplet codons beginning from a specific nucleotide. Depending on where one begins, each DNA strand contains three different reading frames. (See Open reading frame, Overlapping reading frames.)

**Recessive gene:** Characterized as having a phenotype expressed only when both copies of the gene are mutated or missing.

**Recessive(-acting) oncogene, (anti-oncogene):** A single copy of this gene is sufficient to suppress cell proliferation; the loss of both copies of the gene contributes to cancer formation. (See Oncogene.)

**Recessive:** Refers to the member of a pair of alleles that fails to be expressed in the phenotype of the organism when the dominant member is present. Also refers to the phenotype of an individual that has only the recessive allele.

**Recognition sequence (site):** A nucleotide sequence—composed typically of 4, 6, or 8 nucleotides—that is recognized by a restriction endonuclease. Type II enzyrnes cut (and their corresponding modification enzymes methylate) within or very near the recognition sequence.

**Recombinant DNA:** The process of cutting and recombining DNA fragments from different sources as a means to isolate genes or to alter their structure and function.

**Recombination frequency:** The frequency at which crossing over occurs between two chromosomal loci—the probability that two loci will become unlinked during meiosis.

**Refuge:** A refuge is any host plant (non-Bt crop, potatoes, oats, sorghum, and some weeds) not producing Bt proteins or not being treated with conventional Bt formulations. The purpose of the refuge is to supply a source of susceptible insects that may mate with resistant insects of the same species emerging from nearby Bt crop fields. If the refuge is large enough, susceptible insect numbers should outnumber resistant individuals, and in theory, a greater number of the next generation offspring will be susceptible to the Bt crop.

**Regulatory gene:** A gene whose protein controls the activity of other genes or metabolic pathways.

**Relaxed plasmid:** A plasmid that replicates independently of the main bacterial chromosome and is present in 10-500 copies per cell. (See Plasmid.)

**Renature:** The reannealing (hydrogen bonding) of single-stranded DNA and/or RNA to form a duplex molecule.

**Replicon:** A chromosomal region containing the DNA sequences necessary to initiate DNA replication processes.

**Repressor:** A DNA-binding protein in prokaryotes that blocks gene transcription by binding to the operator.

**Resistance:** The evolved ability of a species to survive during a stressful event or set of circumstances.

**Restriction endonuclease (enzyme):** A class of endonucleases that cleaves DNA after recognizing a specific sequence, such as BamH1 (GGATCC), EcoRI (GAATTC), and HindIII (AAGCTT). Type I. Cuts nonspecifically a distance greater than 1000 bp from its recognition sequence and contains both restriction and methylation activities. Type II. Cuts at or near a short, and often symmetrical, recognition sequence. A separate enzyme methylates the same recognition sequence. Type III. Cuts 24-26 bp downstream from a short, asymmetrical recognition sequence. Requires ATP and contains both restriction and methylation activities.

**Restriction enzyme:** A class of enzymes that cut DNA at specific sequences called restriction sites. Restriction enzymes were key to making genetic engineering possible. See RFLP.

**Restriction-fragment-length polymorphism (RFLP):** Differences in nucleotide sequence between alleles at a chromosomal locus result in restriction fragments of varying lengths detected by Southern analysis.

**Retrovirus:** A member of a class of RNA viruses that utilizes the enzyme reverse transcriptase to reverse copy its genome into a DNA intermediate, which integrates into the hostcell chromosome. Many naturally occurring cancers of vertebrate animals are caused by retroviruses.

**Reverse genetics:** Using linkage analysis and polymorphic markers to isolate a disease gene in the absence of a known metabolic defect, then using the DNA sequence of the cloned gene to predict the amino acid sequence of its encoded protein.

**Reverse transcriptase (RNA-dependent DNA polymerase):** An enzyme isolated from retrovirus-infected cells that synthesizes a complementary (c)DNA strand from an RNA template.

**RFLP:** Restriction fragment length polymorphism. RFLPs are generated by digesting DNA with restriction enzymes. The DNA is separated by size by gel electrophoresis. Slight differences in homologous fragments may exist between individuals. These length polymorphisms are RFLPs. See blot.

**Rhizobia:** Bacteria in a symbiotic relationship with leguminous plants that results in nitrogen fixation. See Nitrogen fixation.

**Ribosomal RNA (rRNA):** The RNA component of the ribosome. (See RNA.)

**Ribosome:** Cellular organelle that is the site of protein synthesis during translation. See Organelle, Translation.

**Ribosome-binding site:** The region of an mRNA molecule that binds the ribosome to initiate translation.

**Risk assessment:** Defined as a formalized basis for the objective evaluation of risk in a manner in which assumptions and uncertainties are clearly considered and presented.

**RNA (ribonucleic acid):** An organic acid composed of repeating nucleotide units of adenine, guanine, cytosine, and uracil, whose ribose components are linked by phosphodiester bonds. (See Antisense RNA, Heterogeneous nuclear RNA, Messenger RNA, Ribosomal RNA, RNA polymerase, Small nuclear RNA, Transfer RNA.)

**RNA polymerase:** Transcribes RNA from a DNA template.

**RNA:** Ribonucleic acid. The three basic types are messenger RNA (mRNA), ribosomal RNA (rRNA), and transfer RNA (tRNA).

**Roundup Ready:** A trademark for plants that are genetically engineered to be resistant to the herbicide Roundup (technical name: glyphosate).

**Salmonella:** A genus of rod-shaped, gram-negative bacteria that are a common cause of food poisoning.

**Satellite RNA (viroids):** A small, self-splicing RNA molecule that accompanies several plant viruses, including tobacco ringspot virus.

**Scanning Near Field Optical Microscopy:** A method for observing local optical properties of a surface that can be smaller than the wavelength of the light used. A type of scanning probe microscopy that maps the near-field optical properties of an interface.

**Scanning Probe Microscopy (SPM):** A method for imaging nanoscale features of surfaces by scanning a sensor (probe) over a surface. Near-field effects such as tunneling, van der Waals forces, local fields and more are serially detected at localized points on the surface and used to create an SPM image.

**Scanning Thermal Microscopy (SThM):** A type of scanning probe microscopy that maps the local temperature and thermal conductivity of an interface. A method for observing local temperatures and temperature gradients on a surface. [NTN]

**Scanning Tunneling Microscope (STM):** An instrument able to image conducting surfaces to atomic accuracy; has been used to pin: molecules to a surface. A device in which a sharp conductive tip is moved across a conductive surface close enough to permit a substantial tunneling current (typically a nanometer or less). In a common mode of operation, the voltage is kept constant and the current is monitored and kept constant by controlling the height of the tip above the surface; the result, under favorable conditions, is an atomic-resolution map of the surface reflecting a combination of topography and electronic properties. The STM has been used to manipulate atoms and molecules on surfaces. A type of scanning probe microscopy that maps the local electrical properties of an interface.

**Schwann cell:** Glial cell responsible for forming myelin sheaths in the peripheral nervous system.

**Science:** The process of developing a systematized knowledge of the world through the variation and testing of hypotheses. The pursuit of knowledge and understanding, from the Latin term *scientia*, which means 'knowledge'.

**Selectable marker genes:** Genetic engineers must be able to select cells that have received the transgene from those that have not. Selecting out transgenic cells is done by co-transforming the cells with the transgene plus an additional gene called a selectable marker gene. Selectable marker genes are genes that encode easily detectable traits making transgenic cells discernible from non-transgenic cells. The two most commonly used selectable marker genes encode the traits of herbicide and antibiotic resistance.

**Selectable marker:** A gene whose expression allows one to identify cells that have been transforrned or transfected with a vector containing the marker gene. (See B-Lactamase, Kanr.)

**Self-pollination:** Pollen of one plant is transferred to the female part of the same plant or another plant with the same genetic makeup.

**Semiconservative replication:** During DNA duplication, each strand of a parent DNA molecule is a template for the synthesis of its new complementary strand. Thus, one half of a preexisting DNA molecule is conserved during each round of replication.

**Sequence hypothesis:** Francis Crick's seminal concept that genetic information exists as a linear DNA code; DNA and protein sequence are colinear.

**Sequence:** Noun: the particular order of nucleotides in a DNA or RNA fragment. Verb: to determine the particular order of nucleotides in a strand of DNA.

**Sequence-tagged site (STS):** A unique (single-copy) DNA sequence used as a mapping landmark on a chromosome.

**Sexual reproduction:** The process where two cells (gametes) fuse to form one hybrid, fertilized cell. See Asexual reproduction, Gamete, Hybrid.

**Signal transduction:** The biochemical events that conduct the signal of a hormone or growth factor from the cell exterior, through the cell membrane, and into the cytoplasm. This involves a number of molecules, including receptors, pro- teins, and messengers.

**Site-directed mutagenesis:** The process of introducing spe- cific base-pair mutations into a gene.

**Small nuclear RNA (snRNA):** Short RNA transcripts of 100-300 bp that associate with proteins to form small nuclear ribonucleoprotein particles (snRNPs), which participate in RNA processing. (See RNA.)

**Somatic cell gene therapy:** The repair or replacement of a defective gene within somatic tissue. (See Somatic cell.)

**Somatic cell:** Any nongerm cell that composes the body of an organism and which possesses a set of multiploid chromosomes (diploid in most organisms). (See Gamete, Somatic cell gene therapy.)

**Southern hybridization (Southern blotting):** A procedure in which DNA restriction fragments are transferred from an agarose gel to a nitrocellulose filter, where the denatured DNA is then hybridized to a radioactive probe (blotting). (See Hybridization.)

**Species:** A classification of related organisms that can freely interbreed.

**Spore:** A form taken by certain microbes that enables them to exist in a dormant stage. It is an asexual reproductive cell. See Asexual reproduction, Dormant.

**Stationary phase:** The plateau of the growth curve after log growth, during which cell number remains constant. New cells are produced at the same rate as older cells die. (See Growth phase.)

**Sticky end:** A protruding, single-stranded nucleotide sequence produced when a restriction endonuclease cleaves off center in its recognition sequence. After cutting with restriction enzymes, the ends of DNA fragments are "sticky". They can easily be joined with other fragments that were cut with the same enzyme and so have complementary sticky ends.

**Stringency:** Reaction conditions—notably temperature, salt, and pH—that dictate the annealing of single-stranded DNA/DNA, DNA/RNA, and RNA/RNA hybrids. At high stringency, duplexes form only between strands with perfect one-to-one complementarity; lower stringency allows annealing between strands with some degree of mismatch between bases.

**Stringent plasmid:** A plasmid that only replicates along with the main bacterial chromosome and is present as a single copy, or at most several copies, per cell.

**Structure-functionalism:** The scientific tradition that stresses the relationship between a physical structure and its function, for example, the related disciplines of anatomy and physiology.

**Substantial equivalence:** The concept of substantial equivalence is used as a guide in the safety assessment of genetically modified foods by comparing the novel food to its unmodified counterpart which has a history of safe use. This approach allows regulatory agencies to include in their consideration, the substantial history of information related to foods which have long been safely consumed in the human diet to aid in the identification of potential safety and nutritional issues. *Definition from Health Canada Web site.*

**Subunit vaccine:** A vaccine composed of a purified antigenic determinant that is separated from the virulent organism. (See Vaccine, Enzyme.)

**Supercoiled plasmid:** The predominant in vivo form of plasmid, in which the plasmid is coiled around histone-like proteins. Supporting proteins are stripped away during extraction from the bacterial cell, causing the plasmid molecule to supercoil around itself in vitro. (See Plasmid.)

**Supergene:** A group of neighboring genes on a chromosome that tend to be inherited together and sometimes are functionally related.

**Supernatant:** The soluble liquid &action of a sample after centrifugation or precipitation of insoluble solids.

**Susceptibility:** When individual or species will become ill or possibly die in the face of the stressful event or circumstances.

**Symbiosis:** The close association of two or more dissimilar organisms where both receive an advantage from the association.

**Synapsis:** The pairing of homologous chromosome pairs during prophase of the first meiotic division, when crossing over occurs.

**Taq polymerase:** A heat-stable DNA polymerase isolated from the bacterium Therrnus aquaticus, used in PCR. (See Polymerase.)

**TATA box:** An adenine- and thymine-rich promoter sequence located 25-30 bp upstream of a gene, which is the binding site of RNA polymerase.

**T-DNA (transfer DNA, tumor-DNA):** The transforming region of DNA in the Ti plasmid of Agrobacterium tumefaciens.

**Template:** An RNA or single-stranded DNA molecule upon which a complementary nucleotide strand is synthesized.

**Termination codon:** Any of three mRNA sequences (UGA, UAG, UAA) that do not code for an amino acid and thus signal the end of protein synthesis. Also known as stop codon. (See Codon.)

**Terminator region:** A DNA sequence that signals the end of transcription.

**Tesla [T]:** Unit of magnetic induction: 1T = 1 weber/$m^2$ (also, 1T = $10^4$ gauss).

**Thermistor:** A temperature-measuring device, that contains a resistor or semiconductor whose resistance varies with temperature.

**Thermocouple:** A temperature-measuring device, which contains a pair of end-joined dissimilar conductors in which an electromotive force is developed by thermoelectric effects when the joined ends and the free ends of the conductors are a different temperature.

**Thiol:** An SH group, or a molecule containing one. Also known as a sulfhydryl or mercapto group.

**Threshold:** The smallest input signal that will cause a readable change in the output signal.

**Thrombin :** The enzyme that converts *fibrinogen* to *fibrin* to form a blood clot. Thrombin circulates in the inactive form, prothrombin.

**Thylakoid:** Flattened sac of membrane in a chloroplast that contains pigment and carries out the light-gathering reactions of photosynthesis. Stacks of thylakoids form the grana of chloroplasts.

**Thymidine kinase (tk):** An enzyme that allows a cell to utilize an alternate metabolic pathway for incorporating thymidine into DNA. Used as a selectable marker to identify transfected eukaryotic cells.

**Ti (tumor-inducing) plasmid:** A giant plasmid of Agrobacterium tumefaciens that is responsible for tumor formation in infected plants. Ti plasmids are used as vectors to introduce foreign DNA into plant cells.

**Toxic Substances Control Act (TSCA):** See Environmental Protection Agency.

**Transcapsidation:** The partial of full coating of the nucleic acid of one virus with a coat protein of a differing virus. See Coat protein.

**Transcription:** The process of creating a complementary RNA copy of DNA.

**Tanscription:** The copying of genetic information from DNA to messenger RNA (mRNA).

**Transduction:** The transfer of DNA sequences from one bacterium to another via lysogenic infection by a bacteriophage (transducing phage).

**Transfection:** The uptake and expression of a foreign DNA sequence by cultured eukaryotic cells.

**Transformant:** In prokaryotes, a cell that has been genetically altered through the uptake of foreign DNA. In higher eukaryotes, a cultured cell that has acquired a malignant phenotype. (See Transformation.)

**Transformation:** (1) Transformation is the step in the genetic engineering process where a new gene (transgene) is delivered into the nucleus of a plant cell and inserts into a chromosome where it is passed on to progeny. Transformation means to genetically change a living thing. (2) In prokaryotes, the natural or induced uptake and expression of a foreign DNA sequence—typically a recombinant plasmid in experimental systems. In higher eukaryotes, the conversion of cultured cells to a malignant phenotype—typically through infection by a tumor virus or transfection with an oncogene. (See Transformant, Transformation efficiency.)

**Transformation efficiency:** The number of bacterial cells that uptake and express plasmid DNA divided by the mass of plasmid used (in transformants/microgram). (See Transformation.)

**Transforming oncogene:** A gene that upon transfection converts a previously immortalized cell to the malignant phenotype. (See Oncogene.)

**Transgene:** A foreign gene incorporated by transformation.

**Transgenic:** A genetically engineered plant created through transformation is sometimes referred to as a transgenic.

**Transgenic animal:** Genetically enginnered animal or offspring of genetically engineered animals. The transgenic animal usually contains material from at lease one unrelated organism, such as from a virus, plant, or other animal. See Transgenic.

**Transgenic plant:** Genetically engineered plant or offspring of genetically engineered plants. The transgenic plant usually contains material from at least one unrelated organisms, such as from a virus, animal, or other plant. See Transgenic.

**Transgenic:** An organism in which a foreign DNA gene (a transgene) is incorporated into its genome early in de- velopment. The transgene is present in both somatic and germ cells, is expressed in one or more tissues, and is inherited by offspring in a Mendelian fashion. See Transgenic animal, Transgenic plant.

**Transgenic:** An organism that has incorporated a functional foreign gene through recombinant DNA technology. The novel gene exists in all of its cells and is passed through to progeny. Same as genetically modified organism (GMO).

**Transition-state intermediate:** In a chemical reaction, an unstable and high-energy configuration assumed by reactants on the way to making products. Enzymes are thought to bind and stabilize the transition state, thus lowering the energy of activation needed to drive the reaction to completion.

**Translation:** The process of converting the genetic information of an mRNA on ribosomes into a polypeptide. Transfer RNA molecules carry the appropriate amino acids to the ribosome, where they are joined by peptide bonds.

**Translocation:** The movement or reciprocal exchange of large-chromosomal segments, typically between two different chromosomes.

**Transposition:** The movement of a DNA segment within the genome of an organism.

**Transposon (transposable, or movable genetic element):** A relatively small DNA segment that has the ability to move from one chromosomal position to another.

**tRNA (transfer RNA):** The class of small RNA molecules that transfer amino acids to the ribosome during protein synthesis. See Transfer RNA.

**Trypsin:** A proteolytic enzyme that hydrolyzes peptide bonds on the carboxyl side of the amino acids arginine and lysine.

**TSCA:** The Toxic Substances Control Act. See Environmental Protection Agency.

**U:S: Department of Agriculture:** The U.S. agency responsible for regulation of biotechnology products in plants and animals. The major laws under which the agency has regulatory powers include the Federal Plant Pest Act (PPA), the Federal Seed Act, and the Plant Variety Act (PVA). In addition, the Science and Education (S&E) division has nonregulatory oversight of research activities that the agency funds.

**Upload:** (a) To transfer the consciousness and mental structure of a person from a biological matrix to an electronic or informational matrix (this assumes that the strong AI postulate holds). The term "Downloading" is also sometimes used, mainly to denote transferring the mind to a slower or less spacious matrix. (b) The resulting infomorph person. [The origin of the term is uncertain, but obviously based on the computer technology term 'uploading' (loading data into a mainframe computer).] [AS]

**Vegetal pole:** The end at which most of the yolk is located in an animal egg. The end opposite the animal pole.

**Viral oncogene:** A viral gene that contributes to malignancies in vertebrate hosts. (See Oncogene.)

**Viroid:** A plant pathogen that consists of a naked RNA molecule of approximately 250-350 nucleotides, whose extensive base pairing results in a nearly correct double helix. (See Satellite RNA.)

**Virtual Nanomedicine:** Using VR to perform surgery and other functions inside the body.

**Virtual reality system:** A combination of computer and interface devices (goggles, gloves, etc.) that presents a user with the illusion of being in a three dimensional world of computer-generated objects.

**Virulence:** The degree of ability of an organism to cause disease.

# 1

# Introduction to Theory and Practice of Biotechnology

### Gene Gechnology and the Environment: Regulating Risks and Benefits

Gene technology involves altering the genetic 'blueprint' of living things to make them capable of producing new substances or performing new functions. Genes within a species can be modified, or these basic genetic building block of all organisms can also be transferred between species. Genetically modified organisms or *GMOs* have significant potential to benefit the environment. Using genetically modified plants and animals for agriculture, for example, could help reduce the environmental impact of farming. GMOs also have the potential for harmful impacts. Under the *Gene Technology Act 2000* a licence must be issued before a new GMO is released into the environment. Potential risks must be carefully identified, assessed, and managed to ensure that it does not pose a threat of serious or irreversible damage to the environment. The potential benefits and risks from gene technology, and the ways in which environmental risks are managed, are discussed below. The Department contributed funding to a large CSIRO research project to improve our understanding of the wider ecological impacts of GMOs.

The use of gene technology for managing the environment is at a very early stage, with many of the potential applications still being researched:

- The use of GMOs may provide more effective control of some exotic weeds and feral animal pests on land and in water. The CSIRO, for example, is developing immuno-contraceptive vaccines for foxes, rabbits and mice using gene technology.
- CSIRO and Orica Australia Ltd are using gene technology to develop enzyme products that detoxify pesticide residues, which in Australia would be of particular value to the cotton, horticultural and rice industries.
- Bacteria have been genetically modified as 'bioluminescors' that give off light in response to several chemical pollutants. These are currently being used to measure the presence of some hazardous chemicals in the environment.
- Other genetic sensors that can be used to detect various chemical contaminants are also being trialled. Some can be used to track how pollutants are naturally degrading in ground water.
- It may be possible in the future to include 'sterility' genes in plants and animals imported into Australia or specific regions so that they can breed only in captivity.
- The use of pest-resistant crops could mean a reduced use of insecticides, with less impact on other insects and animals, less pollution of air, waterways and soil, and reduced water and fuel use. The introduction of Bt cotton to Australia, which contains a toxin from the bacterium *Bacillus thuringiensis*, is already resulting in less use of insecticides to control

bollworms.

- The use of genetically modified crops that are tolerant to broad-spectrum herbicides may result in more effective weed control, leading to less overall herbicide use and/or less demand for more narrowly targeted and persistent herbicides.
- Use of some genetically modified herbicide-tolerant crops may result in less cultivation for weed control thereby reducing the degradation of fragile soils.
- More efficient farming may lead to less pressure on the natural environment, including native bushland.

Before introducing new GMOs into the environment, regulators must assess a range of potential risks. Often there are strict conditions associated with the use of a GMO. If the environmental risks cannot be managed appropriately, then the organism is not approved for release. The following are examples of the types of risk considered.

- Immumocontraceptive GMOs could infect non-target native species.
- Overuse or misuse of herbicide-tolerant crops and pastures and the herbicides used on them could lead to herbicide resistant weeds, or sometimes referred to as 'super weeds'. Herbicide-tolerance genes could also transfer to wild populations or weedy relatives of the crop and increase the use of herbicides in non-agricultural areas.
- Insect-resistant crops could lead to insects developing new resistance. In the case of those crops where a bacterial gene is inserted to kill insect herbivores, resistant insects have already developed overseas in response to use of conventional, non-genetically modified products.

- Insect resistant crop plant material, for example from Bt cotton, may impact on soil ecosystems.
- In plants genetically modified to resist viruses, viral 'recombination' could create new viruses, with unknown environmental consequences.
- For genetically modified animals such as pigs or salmon, modifications for faster growth and larger size could transfer to native or feral populations should the GMO escape and interbreed. This could increase the fitness of feral animals, alter population dynamics or breeding patterns and disturb ecosystems.
- Tailoring crops to survive harsh conditions or for better pest or disease control may increase the scale and geographical extent of agriculture and its impact on the natural environment.

For more information see 'Submission by Environment Australia, Inquiry into Primary Producer Access to Gene Technology, House of Representatives Standing Committee on Primary Industries and Regional Services'.

The *Gene Technology Act 2000*, which came into force on 21 June 2001, regulates all dealings with genetically modified organisms in Australia. Under this legislation, licenses must be issued for any 'intentional release' into the environment. Before issuing such a licence, an independent Gene Technology Regulator must prepare a risk assessment and also a risk management plan. The Environment Minister is consulted both before and after these have been prepared. Written public submissions are also invited on the prepared risk assessment and risk management plan. In addition, the Department of the Environment and Heritage administers the *Environment Protection and Biodiversity Conservation Act 1999*. It regulates actions that are likely to have a significant impact on matters of national environmental significance, and also actions on Commonwealth land or actions by the Commonwealth that are likely to have a significant impact on the environment. The Act could be triggered by some dealings with genetically modified organisms.

### Works of Genetic Engineering

Genetic engineering is the process of manually adding new DNA to an organism. The goal is to add one or more new traits that are not already found in that organism. Examples of genetically engineered (transgenic) organisms currently on the market include plants with resistance to some insects, plants that can tolerate herbicides, and crops with modified oil content. To understand how genetic engineering works, there are a few key biology concepts that must be understood DNA is the recipe for life. DNA is a molecule found in the nucleus of every cell and is made up of 4 subunits represented by the letters A, T, G, and C. The order of these subunits in the DNA strand holds a code of information for the cell. Just like the English alphabet makes up words using 26 letters, the genetic language uses 4 letters to spell out the instructions for how to make the proteins an organism will need to grow and live. Small segments of DNA are called genes. Each gene holds the instructions for how to produce a single protein. This can be compared to a recipe for making a food dish. A recipe is a set of instructions for making a single dish. An organism may have thousands of genes. The set of all genes in an organism is called a genome. A genome can be compared to a cookbook of recipes that makes that organism what it is. Every cell of every living organism has a cookbook.

Proteins do the work in cells. They can be part of structures (such as cell walls, organelles, etc). They can regulate reactions that take place in the cell. Or they can serve as enzymes, which speed-up reactions. Everything you see in an organism is either made of proteins or the result of a protein action Small segments of DNA are called genes. Each gene holds the instructions for how to produce a single protein. This can be compared to a recipe for making a food dish. A recipe is a set of instructions for making a single dish. An organism may have thousands of genes. The set of all genes in an organism is called a genome. A genome can be compared to a cookbook of recipes that makes that organism what it is. Every cell of every living organism has a cookbook. DNA is a 'universal language', meaning the genetic code means the same thing in all organisms. It would be like if all cookbooks around the world were written in a single language that everyone knew. This characteristic is critical to the success of genetic engineering. When a gene for a desirable trait is taken from one organism and inserted into another, it gives the 'recipient' organism the ability to express that same trait. Genetic engineering, also called transformation, works by physically removing a gene from one organism and inserting it into another, giving it the ability to express the trait encoded by that gene. It is like taking a single recipe out of a cookbook and placing it into another cookbook.

**The Process: Once a Goal is in Mind...**

1. First, find an organism that naturally contains the desired trait.
2. The DNA is extracted from that organism. This is like taking out the entire cookbook.
3. The one desired gene (recipe) must be located and copied from thousands of genes that were extracted. This is called gene cloning.
4. The gene may be modified slightly to work in a

more desirable way once inside the recipient organism.

5. The new gene(s), called a transgene is delivered into cells of the recipient organism. This is called transformation. The most common transformation technique uses a bacteria that naturally genetically engineer plants with its own DNA. The transgene is inserted into the bacteria, which then delivers it into cells of the organism being engineered. Another technique, called the gene gun method, shoots microscopic gold particles coated with copies of the transgene into cells of the recipient organism. With either technique, genetic engineers have no control over where or if the transgene inserts into the genome. As a result, it takes hundreds of attempts to achieve just a few transgenic organisms.
6. Once a transgenic organism has been created, traditional breeding is used to improve the characteristics of the final product. So genetic engineering does not eliminate the need for traditional breeding. It is simply a way to add new traits to the pool.

Although the goal of both genetic engineering and traditional plant breeding is to improve an organism's traits, there are some key differences between them. While genetic engineering manually moves genes from one organism to another, traditional breeding moves genes through mating, or crossing, the organisms in hopes of obtaining offspring with the desired combination of traits. Using the recipe analogy, traditional breeding is like taking two cookbooks and combining every other recipe from each into one cookbook. The product is a new cookbook with half of the recipes from each original book. Therefore, half of the genes in the offspring of a cross come from each parent. Traditional breeding is effective in improving traits, however, when compared with genetic engineering, it does have disadvantages. Since breeding relies on the ability to mate two organisms to move genes, trait improvement is basically limited to those traits that already exist within that species. Genetic engineering, on the other hand, physically removes the genes from one organism and places them into the other. This eliminates the need for mating and allows the movement of genes between organisms of any species. Therefore, the potential traits that can be used are virtually unlimited. Breeding is also less precise than genetic engineering. In breeding, half of the genes from each parent are passed on to the offspring. This may include many undesirable genes for traits that are not wanted in the new organism. Genetic engineering, however, allows for the movement of a single, or a few, genes.

## Biotechnology: Applications and Theories

Major changes in the field of biology have taken place over the last few decades. Advances in biotechnology and accomplishments such as the mapping of the entire human genome and the genomes of many other organisms are producing a better understanding of biological systems and living organisms. Large amounts of biological information and data have been and continue to be generated through this research. Bioinformatics arose from the need to organize, store, and analyse this information.

Bioinformatics is the use of computer software and programs to organize, store and analyse biological information and data to better understand biological systems. It is the field of science where biology, computer science and information technology converge as a single discipline. Researchers use advanced computer and statistical techniques to wade through and analyse large amounts of biological data created through modern biotechnologies referred to as the "omics" technologies – such as genomics and proteomics. These "omics" technologies focus on different aspects of biological systems. For example:

- Genomics is the analysis of the entire genetic make-up of an organism. It includes a study of the functions of genes in cells, organs and organisms.
- Proteomics focusses on the total protein make-up of a cell at any given time.

These "omics" technologies generate massive amounts of biological data through which it would be impossible to navigate without the use of computer systems. The data includes sequences of amino acids and nucleotides that underlie genes and proteins. The goal of bioinformatics is to translate the complex data gathered into usable knowledge. Bioinformatics typically includes three main areas:

- Developing new statistics and data to be studied and used to assess relationships among the different sets of biological data.
- Analysing and interpreting the various types of data, which include nucleotide and amino acid sequences and the properties of various proteins.
- Developing and implementing tools to enable efficient access and management of different types of pertinent information.

The field of bioinformatics emerged in the early 1980s with the creation of the GenBank database. GenBank, started by the US Department of Energy, stores DNA sequence information obtained from a wide range of organisms. In the early days, GenBank was a small scale operation with a roomful of technicians sitting at keyboards entering the

DNA sequence information published in academic journals.

The advent of the Internet allowed researchers to access data in GenBank from all over the world for free. The DNA sequence data in GenBank grew rapidly with the emergence of highly sophisticated gene sequencing tools. Private companies joining the sequencing race with parallel projects created huge databases of their own.

Several services emerged as the need for access to bioinformatics increased. The two most significant were the European Molecular Biology Network (EMBnet) and the United States National Center for Biotechnology Information (NCBIO). In Canada, the Canada Institute for Scientific and Technical Information (CISTI) of the National Research Council of Canada (NRC) operated a bioinformatics server, Molecular Biology Database Service (MBDS), from 1987 to 1996. The Canadian Bioinformatics Resource (CBR) was created in 1995 to replace MBDS in providing bioinformatics resources and services to Canadian researchers.

### *Bioinformatics Applications*

Most bioinformatics tools currently available deal with the structural and functional aspects of genes and proteins – looking at what genes and proteins look like, where they are located and what they do. Applications of bioinformatics allow researchers to accomplish the following:

- Compare mapped genomes of any species to identify similarities and differences among organisms.
- Researchers can access genetic information from around the world. Various searchable databases, many of which are on the Internet, are used to identify DNA sequences of identified genes.

- Databases also allow quick access to information about proteins. Researchers retrieve a sequence using a protein name, or retrieve a protein using the amino acid sequence.
- Sequence translation programs enable users to analyse nucleotide sequences. The programs convert the DNA sequences into protein sequences or protein sequences into their complementary DNA (cDNA).
- Sequencing tools are used by bioinformatists to find out more about the characteristics of genetic material, including proteins. Once the protein sequences are analysed, the researchers can use software programs to search databases to find similar sequences. Other programs allow researchers to look at the three-dimensional shape of the proteins and nucleotides. Since shape is a critical determinant of function, these programs allow researchers to gain clues for identifying the roles of the sequences.

### *Bioinformatics Tools*

Data produced through research from all over the world is collected and organized in databases specialized in individual subjects. Computational tools are used to efficiently analyse this data. Many areas of bioinformatics focus on predicting the biological functions of genes, proteins, and their parts based on the structural data compiled. Examples of commonly used tools include:

- BLAST (Basic Local Alignment Search Tool) — This tool allows researchers to identify similarities between their own nucleotide or protein sequence with those in the public databases. It is used to identify whether a given sequence is new, similar to a known sequence (e.g. other proteins), or contains specific protein characteristics which may give a clue to the possible role of the sequence.

- OMIM (Online Mendelian Inheritance in Man) — This database contains information about human genes and genetic disorders already mapped in the human genome. Typically, the information contained includes a summary of each disease, including symptoms, genetic mapping information, related references, and a direct link to the DNA sequence in GenBank if the gene responsible for the disease has been cloned. It also contains mapping information on genes not yet linked to diseases.

GeneMap98 — This database lists genes mapped in the human genome. It is helpful to researchers who are looking to identify genes for diseases with simple traits or symptoms since the map can display genes within a location defined by the user.

Bioinformatics is an interdisciplinary field that addresses biological problems using computational techniques. The field is also often referred to as computational biology. It plays a key role in various areas like functional genomics, structural genomics, and proteomics amongst others, and forms a key component in biotechnology and pharmaceutical sector. Bioinformatics and computational biology involve the use of techniques from applied mathematics, informatics, statistics, and computer science to solve biological problems. Research in computational biology often overlaps with systems biology. Major research efforts in the field include sequence alignment, gene finding, genome assembly, protein structure alignment, protein structure prediction, prediction of gene expression and protein-protein interactions, and the modeling of evolution. The terms *bioinformatics* and *computational biology* are often used interchangeably, although the former typically focuses on algorithm development and specific computational methods, while the latter focuses more on hypothesis testing and discovery in the biological domain. Although this distinction is used by National Institutes of Health in their working definitions of Bioinformatics and Computational Biology, it is clear that there is a tight coupling of developments and knowledge between the more hypothesis-driven research in computational biology and technique-driven research in bioinformatics. Computational biology also includes lesser-known but equally important sub-disciplines such as computational biochemistry and computational biophysics. A common thread in projects in bioinformatics and computational biology is the use of mathematical tools to extract useful information from noisy data produced by high-throughput biological techniques such as genomics (The field of data mining overlaps with computational biology in this regard). A representative problem in bioinformatics is the assembly of high-quality DNA sequences from fragmentary "shotgun" DNA sequencing, while in computational biology, a representative problem might be statistical testing of a hypothesis of common gene regulation using data from mRNA microarrays or mass spectrometry.

In the context of genomics, annotation is the process of marking the genes and other biological features in a DNA sequence. The first genome annotation software system was designed in 1995 by Owen White, who was part of the team that sequenced and analyzed the first genome of a free-living organism to be decoded, the bacterium Haemophilus influenzae. Dr. White built a software system to find the genes (places in the DNA sequence that encode a protein), the transfer RNA, and other features, and to make initial assignments of function to those genes. Most current genome annotation systems work similarly, but the programs available for analysis of genomic DNA are constantly changing and improving.

Evolutionary biology is the study of the origin and descent of species, as well as their change over time. Informatics has assisted evolutionary biologists in several key ways; it has enabled researchers to:

- trace the evolution of a large number of organisms by measuring changes in their DNA, rather than through physical taxonomy or physiological observations alone,
- more recently, compare entire genomes, which permits the study of more complex evolutionary events, such as gene duplication, lateral gene transfer, and the prediction of bacterial speciation factors,
- build complex computational models of populations to predict the outcome of the system over time
- track and share information on an increasingly large number of species and organisms

Future work endeavors to reconstruct the now more complex tree of life. The area of research within computer science that uses genetic algorithms is sometimes confused with computational evolutionary biology. Work in this area involves using specialized computer software to improve equations, algorithms, or integrated circuit designs. It is inspired by evolutionary principles such as replication, diversification through recombination or mutation, fitness, survival through selection or culling, and iteration, collectively called a Darwinian machine or Darwinian ratchet.

Biodiversity of an ecosystem might be defined as the total genomic complement of a particular environment, from all of the species present, whether it is a biofilm in an abandoned mine, a drop of sea water, a scoop of soil, or the entire biosphere of the planet Earth. Databases are used to collect the species names, descriptions, distributions, genetic information, status and size of populations, habitat needs, and how each organism interacts with other species. Specialized software programs are used to find, visualize, and analyze the information, and most importantly, communicate it to other people. Computer simulations model such things as population dynamics, or calculate the cumulative genetic health of a breeding pool (in agriculture) or endangered population (in conservation). One very exciting potential of this field is that entire DNA sequences, or genomes of endangered species can be preserved, allowing the results of Nature's genetic experiment to be remembered *in silico*, and possibly reused in the future, even if that species is eventually lost.

The expression of many genes can be determined by measuring mRNA levels with multiple techniques including microarrays, expressed cDNA sequence tag (EST) sequencing, serial analysis of gene expression (SAGE) tag sequencing, massively parallel signature sequencing (MPSS), or various applications of multiplexed in-situ hybridization. All of these techniques are extremely noise-prone and/or subject to bias in the biological measurement, and a major research area in computational biology involves developing statistical tools to separate signal from noise in high-throughput gene expression studies. Such studies are often used to determine the genes implicated in a disorder: one might compare microarray data from cancerous epithelial cells to data from non-cancerous cells to determine the transcripts that are up-regulated and down-regulated in a particular population of cancer cells.

Regulation is the complex orchestra of events starting with an extra-cellular signal and ultimately leading to the increase or decrease in the activity of one or more protein molecules. Bioinformatics techniques have been applied to explore various steps in this process. For example, promoter analysis involves the elucidation and study of sequence motifs in the genomic region surround the coding region of a gene. These motifs influence the extent to which that region is transcribed into mRNA. Expression data can be used to infer gene regulation: one might compare microarray data from a wide variety of states of an organism to form hypotheses about the genes involved in each state. In a single-cell organism, one might compare stages of the cell cycle, along with various stress conditions (heat shock, starvation, etc.). One can then apply clustering algorithms to that expression data to determine which genes are co-expressed. Further analysis could take a variety of directions: one 2004 study analyzed the promoter sequences of co-expressed (clustered together) genes to find common regulatory elements and used machine learning techniques to identify the promoter elements involved in regulating each cluster.

Protein microarrays and high throughput (HT) mass spectrometry (MS) can provide a snapshot of the proteins present in a biological sample. Bioinformatics is very much involved in making sense of protein microarray and HT MS data; the former approach faces similar problems as with microarrays targeted at mRNA, the latter involves the problem of matching large amounts of mass data against predicted masses from protein sequence databases, and the complicated statistical analysis of samples where multiple, but incomplete peptides from each protein are detected.

Massive sequencing efforts are currently underway to identify point mutations in a variety of genes in cancer. The sheer volume of data produced requires automated systems to read sequence data, and to compare the sequencing results to the known sequence of the human genome, including known germline polymorphisms. Oligonucleotide microarrays, including comparative genomic hybridization and single nucleotide polymorphism arrays, able to probe simultaneously up to several hundred thousand sites throughout the genome are being used to identify chromosomal gains and losses in cancer. Hidden Markov model and change-point analysis methods are being developed to infer real copy number changes from often-noisy data. Further informatics approaches are being developed to understand the implications of lesions found to be recurrent across many tumors. Some modern tools (e.g. Quantum 3.1) provide tool for changing the protein sequence at specific sites through alterations to its amino acids and predict changes in the bioactivity after mutations.

Protein structure prediction is another important application of bioinformatics. The amino acid sequence of a protein, the so-called *primary structure*, can be easily determined from the sequence on the gene that codes for it. In the vast majority of cases, these primary structures uniquely determine a structure in its native environment. (Of course, there are exceptions, such as the bovine spongiform encephalopathy - aka Mad Cow Disease - prion.) Knowledge of this structure is vital in understanding the function of the protein. For lack of better terms, structural information is usually classified as one of *secondary*, *tertiary* and *quaternary* structures. A viable general solution to such predictions remains an open problem. As of now, most efforts have been directed towards heuristics that work most of the time. One of the key ideas in bioinformatics research is the notion of homology. In the genomic branch of bioinformatics, homology is used to predict the function of a gene: if the sequence of gene *A*, whose function is known, is homologous to the sequence of gene *B*, whose function is unknown, one could infer that B may share A's function. In the structural branch of bioinformatics homology is used to determine which parts of the protein are important in structure formation and interaction with other proteins. In a technique called homology modeling, this information is used to predict the structure of a protein once the structure of a homologous protein is known. This currently remains the only way to predict protein structures reliably. One example of this is the similar protein homology between hemoglobin in humans and the hemoglobin in legumes (leg hemoglobin). Both serve the same purpose of transporting oxygen in both organisms. Though both of these proteins have completely different amino acid sequences, their protein structures are virtually identical, which reflects their near identical purposes. Other techniques for predicting protein structure include protein threading and *de novo* (from scratch) physics-based modeling.

The core of comparative genome analysis is the establishment of the correspondence between genes (orthology analysis) or other genomic features in different organisms. It is these inter-genomic maps that make it possible to trace the evolutionary processes responsible for the divergence of two genomes. A multitude of evolutionary events acting at various organizational levels shape genome evolution. At the lowest level, point mutations affect individual nucleotides. At a higher level, large chromosomal segments undergo duplication, lateral transfer, inversion, transposition, deletion and insertion. Ultimately, whole genomes are involved in processes of hybridization, polyploidization and endosymbiosis, often leading to rapid speciation. The complexity of genome evolution poses many exciting challenges to developers of mathematical models and algorithms, who have recourse to a spectra of algorithmic, statistical and mathematical techniques, ranging from exact, heuristics, fixed parameter and approximation algorithms for problems based on parsimony models to Markov Chain Monte Carlo algorithms for Bayesian analysis of problems based on probabilistic models. Many of these studies are based on the homology detection and protein families' computation.

Systems biology involves the use of computer simulations of cellular subsystems (such as the networks of metabolites and enzymes which comprise metabolism, signal transduction pathways and gene regulatory networks) to both analyze and visualize the complex connections of these cellular processes. Artificial life or virtual evolution attempts to understand evolutionary processes via the computer simulation of simple (artificial) life forms.

Computational technologies are also used to accelerate or fully automate the processing, quantification and analysis of large amounts of high-information-content Biomedical imagery. Modern image analysis systems augment the observers' ability to make measurements from a large or complex set of images, by improving accuracy, objectivity, or speed. A fully developed analysis system may completely replace the observer. While these systems are not unique to biology related imagery, their application to biologic problems continue to provide unique challenges and solutions, placing several imagery application under the umbrella of Bioinformatics. These systems are in the process of becoming more important for both diagnostics and research. Some examples:

- high-throughput and high-fidelity quantification and sub-cellular localization (high-content screening, cytohistopathology)
- morphometrics are used to analyze pictures of embryos to track and to predict the fate of cell clusters during morphogenesis
- clinical image analysis and visualization

- determine the real-time air-flow patterns in breathing lungs of living individuals before and during challenge
- quantify occlusion size in real-time imagery from the development of and recovery during arterial injury
- making behavioral observations from extended video recordings of laboratory animals
- infrared measurements for metabolic activity determination

### *Biotechnology Software Tools*

The computational biology tool best-known among biologists is probably BLAST, an algorithm for searching large sequence (protein, DNA) databases. NCBI provides a popular implementation that searches their massive sequence databases. Bioinformatic meta search engines (Entrez, Bioinformatic Harvester) help finding relevant information from several databases. There are also free Web-based software designed for structural bioinformatics such as STING. Computer scripting languages such as Perl and Python are often used to interface with biological databases and parse output from bioinformatics programs. Communities of bioinformatics programmers have set up free/open source projects such as EMBOSS, Bioconductor, BioPerl, BioLinux, BioPython, BioRuby, and BioJava which develop and distribute shared programming tools and objects (as program modules) that make bioinformatics easier. An integrated software workbench consisting of many free/open source tools described above and many others is known as VigyaanCD. Taverna an open-source bioinformatics workbench that utilizes a workflow model of experimental design. Taverna is included as part of the myGRID package of e-science software. Quantum 3.1 is an example of the bioinformatics post-QSAR technology applying quantum and molecular physics instead of statistical methods. Genevestigator is an example of how large-scale gene expression microarray data is used to predict gene function based on contextual information. More recently, SOAP-based interfaces have been developed for a wide variety of bioinformatics applications such as blast, fasta, EMBOSS, clustalw, t-coffee, MUSCLE and many others. These are available from the EBI at EBI Web Services.

# 2

# Theories of Biotechnology: An Overview

Many people do not understand current ideas about evolution. The following is a brief summary of the modern consensus among evolutionary biologists. The idea that life on Earth has evolved was widely discussed in Europe in the late 1700's and the early part of the last century. In 1859 Charles Darwin supplied a mechanism, namely natural selection, that could explain how evolution occurs. Darwin's theory of natural selection helped to convince most people that life has evolved and this point has not been seriously challenged in the past one hundred and thirty years. It is important to note that Darwin's book "The Origin of Species by Means of Natural Selection" did two things. It summarized all of the evidence in favor of the idea that all organisms have descended with modification from a common ancestor, and thus built a strong case for evolution. In addition Darwin advocated natural selection as a mechanism of evolution. Biologists no longer question whether evolution has occurred or is occurring. That part of Darwin's book is now considered to be so overwhelmingly demonstrated that is is often referred to as the FACT of evolution. However, the MECHANISM of evolution is still debated. We have learned much since Darwin's time and it is no longer appropriate to claim that evolutionary biologists believe that Darwin's theory of Natural

Selection is the best theory of the mechanism of evolution. I can understand why this point may not be appreciated by the average non-scientist because natural selection is easy to understand at a superficial level. It has been widely promoted in the popular press and the image of "survival of the fittest" is too powerful and too convenient.

During the first part of this century the incorporation of genetics and population biology into studies of evolution led to a Neo-Darwinian theory of evolution that recognized the importance of mutation and variation within a population. Natural selection then became a process that altered the frequency of genes in a population and this defined evolution. This point of view held sway for many decades but more recently the classic Neo-Darwinian view has been replaced by a new concept which includes several other mechanisms in addition to natural selection. Current ideas on evolution are usually referred to as the Modern Synthesis which is described by Futuyma;

"The major tenets of the evolutionary synthesis, then, were that populations contain genetic variation that arises by random (ie. not adaptively directed) mutation and recombination; that populations evolve by changes in gene frequency brought about by random genetic drift, gene flow, and especially natural selection; that most adaptive genetic variants have individually slight phenotypic effects so that phenotypic changes are gradual (although some alleles with discrete effects may be advantageous, as in certain color polymorphisms); that diversification comes about by speciation, which normally entails the gradual evolution of reproductive isolation among populations; and that these processes, continued for sufficiently long, give rise to changes of such great magnitude as to warrant the designation of higher taxonomic levels (genera, families, and so forth)." — Futuyma, D.J. in *Evolutionary Biology*.

This description would be incomprehensible to Darwin since he was unaware of genes and genetic drift. The modern theory of the mechanism of evolution differs from Darwinism in three important respects:

1. It recognizes several mechanisms of evolution in addition to natural selection. One of these, random genetic drift, may be as important as natural selection.
2. It recognizes that characteristics are inherited as discrete entities called genes. Variation within a population is due to the presence of multiple alleles of a gene.
3. It postulates that speciation is (usually) due to the gradual accumulation of small genetic changes. This is equivalent to saying that macroevolution is simply a lot of microevolution.

In other words, the Modern Synthesis is a theory about how evolution works at the level of genes, phenotypes, and populations whereas Darwinism was concerned mainly with organisms, speciation and individuals. This is a major paradigm shift and those who fail to appreciate it find themselves out of step with the thinking of evolutionary biologists. Many instances of such confusion can be seen here in the newsgroups, in the popular press, and in the writings of anti-evolutionists. The major controversy among evolutionists today concerns the validity of point #3 (above). The are many who believe that the fossil record at any one site does not show gradual change but instead long periods of stasis followed by rapid speciation. This model is referred to as Punctuated Equilibrium and it is widely accepted as true, at least in some cases. The debate is over the relative contributions of gradual versus punctuated change, the average size of the punctuations, and the mechanism. To a large extent the debate is over the use of terms and definitions, not over fundamentals. No new mechanisms of evolution are needed to explain the model. Some scientists continue to refer to modern thought in evolution as Neo-Darwinian. In some cases these scientists do not understand that the field has changed but in other cases they are referring to what I have called the Modern Synthesis, only they have retained the old name.

## Genetic Exploration: An Expanded Perspective

Geneticists, generally trained within some form of evolutionary paradigm, assume it as the underlying context within which they search and interpret. This paper, written by a generalist for scientists, is submitted as a brief to inform the reader of an archaeological paradigm, developed over a century and a half, which renders outmoded, corrects and subsumes both the Darwinian evolutionary and creationist models, as related to human history, and provides an expanded, fresh perspective on DNA research. The most fundamental, critical obstacle blocking interpretive and explanatory progress in the sciences – as well as our educational, political and social processes — is our lack of a consensual definition of what a generic human being is, what it is to be generically human. Theological, philosophical, scientific and new age definitions rattle around in a criteria vacuum. The creationist-evolutionist conflict is a single, classic example of how effectively we have been Babel-factored into ineffective argument without closure for decades. The thesis of the Sumerian scholar, Zecharia Sitchin (1976), provides a robust paradigm which affords the resolution of this, objectively astounding, enervating and divisive problem at the heart of western culture. It frees the discussions of the science, ethics, and philosophy involved from the recycled contexts, criteria and syllogistic gambits which have stultified the western intellect for so long. Its startling anthropological ramifications provide, not only a generic definition of what a human is but, through a recovered and restored history of our bicameral genome's initiation and evolution, no less than a key to the decipherment and interpretation of the human genome itself. Correlated with the current data from the genome projects, general genetics, cloning, and related areas it, further, furnishes vital clues to an explanation of the anomalous fact of our presenting with four thousand plus and counting genetic diseases, a possible reason for so-called "junk" genes, and indicators as to why we should possibly expect to find certain suppressed genes. A potentially vital and startling contribution to human history by the science of genetics, as a spin off of the current ongoing genome research, is suggested.

### A Synopsis of Sitchin's Archaeological Thesis

Zecharia Sitchin is one of two hundred Sumerian and Semitic scholars on the planet who can read a clay tablet like you and I can read this text. He published his first work, The Twelfth Planet, in 1976 and, subsequently, another six volumes of the Earth Chronicle Series. From his knowledge of Sumerian, the Semitic and other ancient languages and his research on the same archaeological and Biblical data and materials studied by all archaeologists, acquired over the last one hundred and fifty years, he put forth a radical thesis: the transcultural "gods" known to all the ancient cultures (Sumerian: Anunnaki, meaning "those who came down from the heavens") were not mythological but real flesh and blood humanoid aliens, very much like humans, who had come here from the tenth planet in our solar system, Planet X in the popular press, called Nibiru by the Sumerians, colonizing Earth some 432,000 years ago. They subsequently genetically engineered our species, 200,000 years ago, originally as slave animals to work in their gold mines, by crossing their own genes with those of Homo

Erectus. We became subservient partners with them and, eventually, they phased off the planet leaving our species on our own.

Before one rejects this thesis out of hand, it is well to understand that it is backed up by highly coherent, well researched, well developed archaeological data, documents, history and some two million items of recovered artifacts. A key factor is our having reached a level of technological advancement which allows us to understand high science and technologies present in the past which, until only very recently, could only seem fantastic – particularly genetic engineering. As is the case with scientists in general, archaeologists can adjust to internal and incidental shifts in their consensual context but, when the paradigm itself is threatened with replacement, there inevitably occurs a great deal of very unscientific chest drumming and incisor display. Obviously, the obstacle here is the word "alien". Consider that the cardinal reason why the alien topic has been considered the fringe "belief" of unbalanced personalities for so long has been the relentless promotion of that idea by the government for half a century. The Brookings Institution (think tank) study (early '50's, commissioned by the government) concluded that revelation of the alien presence would cause serious disruption in the fundamentalist religions and society and recommended against. But extra-solar system aliens are not our focus here.

The Anunnaki, as depicted in hundreds of carvings, reliefs, drawings, and descriptions in the recovered documents, are quite like us, male and female, generally taller and huskier than humans in those times, from within our solar system, at home in our atmosphere, radically different than the alien types allegedly from other solar systems. If we are half Anunnaki, that should not be so surprising. At the time they were on the planet, they apparently possessed only ballistic rocketry for interplanetary travel, no anti-gravitic technology yet developed.

*Objections to this Thesis:* Those trained in the sciences, accustomed to carefully crafted protocols and rigorous proof will, no doubt, question this paradigm in several ways. Quite reasonably, it may be asked why the Sitchin paradigm has not been embraced by the geneticists' academic counterparts in archaeology, anthropology, and paleontology if the evidence is as voluminous and robust as claimed. It may also be asked, Why has this not been the known and taught history of our kind through all the millennium, why should there have been suppression and perversion of this knowledge? These questions are best answered in historical context and perspective.

There was literally no such thing as the discipline known as Archaeology in Western culture until the 1800's. The Roman Church controlled and determined the view of the past. The scholastic world, dominated by the Church, followed docilely. Not until paleontological findings of millions of years forced that view to be reevaluated and Schliemann, a wealthy German merchant, refusing to believe that the ancient cities and peoples were legend, dug up several stages of the city of Troy, was a window into the past opened and the mythic view questioned. Scientific Archaeology, as we know it, came into existence only when academics reluctantly had to acknowledge the past being dug up and collected by amateurs in the Middle East. Archaeologists promptly came to be mistrusted and hated by the religious institutions who feared revelations that would contradict their teachings and the history of the Old Testament.

Most in the scientific world are familiar with the scientist, Galileo, having to capitulate to the Inquisition to save his own life, dying while under house arrest for holding to a heliocentric view of the solar system, claiming to see planets through his telescope. Fewer are aware of the fact that the monk, Jordano Bruno, was burnt at the stake in Rome, through the solicitousness of the Roman Church, only thirty six years before the founding of Harvard University, for holding to the Copernican view and claiming that there had to be other planets and other civilizations in the cosmos.

No one in America, literally, had a Doctorate, at the time of the founding of Harvard University, in 1636 by John Harvard. Increase Mather, a president of Harvard, as a Dissenter, was ineligible for a Doctorate from any English university because all of them were controlled by the Church. Fourteen years after the founding of Harvard, Bishop James Uusher published his "Annales Vertis et Novi Testamenti" dating the beginning of the world at 4004 B.C. One could be condemned as a heretic for contraverting this doctrine by decree of the Church in 1654 and the stricture was not removed until 1952 (!) by Pope Pius XII. The arithmetical wonder of this fact is that was only 48 years ago. Consider that almost everything written in this paper would have been branded as "heresy" and who knows how DNA research would have been branded only a short time ago.

As late as 1906, the Egyptian Exploration Fund, whose charter stated it was set up to promote archaeologists whose work would reinforce the Old and New Testament, refused to publish the discovery of an Anunnaki gold processing plant on Mount Horeb in the Sinai by Sir Flinders Petrie, the most distinguished in his field at the time. When he published privately they pulled his funding, had the book expunged from the publisher's records and the British Library never cataloged the work — on one of the most important discoveries in archaeology.(Sir Laurence Gardner: Bloodline Of The Holy Grail, Genesis Of the Grail Kings )

The mythic interpretation has been promulgated by religions because to recognize the Anunnaki as real would be to open the door to a radical reinterpretation of the entire phenomenon of religion and put into question the real identity of the very deity at center of their belief system. To relegate all the Anunnaki "gods" and their deeds — except Enlil/Jehovah/Yahweh, the God of the Judeo-Christian religions, who was sublimated out of that category — to fictional, mythic, unreal status was supremely effective to this end. It is through this millenniums-old tradition of suppression, mythization and manipulative control that the character, content and interpretation in the academic arena has been set and remains, largely, even to this day. Very few are going to make a rubber burning one-eighty over their Ph.D. thesis in Mythology. In Archaeology, Anthropology, and Paleontology, tenure tetanus prevents most from "going first" to admit they have been wrong in a turn as significant and profoundly revolutionary, perhaps even more so, than the Darwinian shock. Collegiate colitis is all too frequent just from contemplating having to contradict oneself in front of students. Peer pressure finishes off all but the isolated courageous academic or scientist here and there. Thomas Kuhn (The Structure of Scientific Revolutions) is proven correct again. If you want that Ph.D. diploma you will repeat what you know you are supposed to believe and say. If you want that job, that funding... The reason becomes apparent why the academic arena has attempted to ignore this forbidden archaeology and anthropology, and why a few more ruthless of the academic power elite have tried to get Sitchin quick and nasty ad hominem just outside the fact forum.

To this day, Anglican clergy still sit on the boards of English universities. In America, the Constitution is only a negative restraint on religious authoritarian domination

Our schools of higher "learning" still parade the trappings of the medieval university on ceremonial occasions and, unfortunately, all too often still in their limitation of discussion to approved subjects. That the influence of religion on science is still virulent is manifest in the deference shown by politicians to religionists in decisions in matters of genetic research.

The ordinary scientist, although perhaps open and cognizant of the serious scientific objections that have been raised against classic Darwinian theory, is reluctant to question or reassess the evolutionary paradigm because of the erroneous fear that creationism would be the only alternative. In effect, the acceptance of the new paradigm is more difficult because it supercedes and corrects both the creationist and the Darwinian models. The Creationists were only half wrong and the Evolutionists only half right: there was a creation event but it was a genetic engineering process; there is an evolutionary process but it was not linear from the time the lightning hit the mud to us as the pinnacle of creation. It was interrupted in our regard by the Anunnaki for their own practical purposes. We shall be compelled to introduce an additional category: a genetically synthesized species.( And, if a voice out there is heard to say "Well, thank God for Darwin regardless of how imperfect his work was.." many of us will still not catch the irony in that statement....)

In our time, the position of the genetic specialist can become quite personally disconcerting. Having been trained as an undergraduate to uncritically accept some variation on the Darwinian evolutionary theme, tending to look to fellow scientists, particularly the archaeologists, biologists, and anthropologists for answers to the "big" questions, conditioned, if only subliminally, by religio-cultural tradition, the geneticist finds it difficult enough to decipher the genome much less consider a radically revolutionary paradigm forcing a complete rethinking of the planet. The purpose of this paper is to respectfully facilitate just that process for the geneticist because the new paradigm contains vital clues and vectors directly related to the geneticist's quest.

*The Demise of the Mythological Explanation of the Gods:* Those still holding the mythological interpretation are in an unenviable position. They find themselves having to hold that the same citizens of the first civilizations who, they claim, invented — or hallucinated — the gods through their primitive imaginings and naive proto-scientific projections of personality on the great forces of nature were the same primitives who somehow could build the stupendous Giza pyramid; somehow quarry, cut and move into place the one thousand ton stones of the Baalbek "temple" (rocket platform) which even our modern technology cannot begin to lift; somehow know the great precessional cycle of the heavens, the existence of all ten of the planets of our solar system and how our solar system was formed (Enuma Elish document). The physical locations of the ancient legends as mythic has been gradually disproven beginning with the work of Schliemann and completed with the re-discovery of all the ancient cities and centers on all continents. The events involving the "gods" of ancient times and the technology attributed to them, previously considered as myth and naive legend, have been gradually proven to be real, beginning with the discovery of documents from the most ancient cities and reinforced with the discovery of the great library of Ashurbanipal at Nineveh and the gradual accumulation of some two million pieces of artifact and document confirming those events in great detail.The discovery of ooparts, (high tech tools, toys, artifacts, and technology along with documentation of advanced scientific knowledge ostensibly out of place in time) along with astronomical knowledge of the entire solar system beyond our current level has reinforced the negation of the mythic interpretation.

The interpretation of the "gods", the Anunnaki themselves, as mythic, unreal beings (academic mythologists) and Jungian archetypes (Joseph Campbell) and the relegation of them to schizophrenic hallucination (Julian Jaynes) has been gradually disproven beginning with the acknowledgment of the reality of the events attributed to them; the discounting of the arguments for their unreality due to the seeming fantastic deeds attributed to them by the development of our technology (rockets, lasers, radio communication, genetic engineering, atomic and particle beam weapons) that duplicate those feats. This interpretation has been implemented further by the exploration and questions posed early by the German school, pioneers like von Daniken and then completed by the comprehensive and brilliant demonstrations of Sitchin and Sir Laurence Gardner, the English historian and genealogist.

Now that we have begun by walking on the moon and exploring the solar system and have probes going starward, the possibility of an alien civilization coming here is taken for granted and one coming here from inside our solar system trivial, rather than unreal myth. "Mythinformation", after two hundred years of failure, although still hiding behind tenure in the university, is a dead issue.

*The Essence of the Sitchin Paradigm:* Working from the same archaeological discoveries, artifacts, and recovered records as archaeologists and linguists have for two hundred years, Sitchin propounds – proves, in the opinion of this author —that the Anunnaki (Sumerian: "those who came down from the heavens"; Old testament Hebrew, Anakeim, Nefilim, Elohim; Egyptian: Neter), an advanced civilization from the tenth planet in our solar system, splashed down in the Persian gulf area around 432,000 years ago, colonized the planet, with the purpose of obtaining large quantities of gold. Some 250,000 years ago, the recovered documents tell us, their lower echelon miners rebelled against the conditions in the mines and the Anunnaki directorate decided to create a creature to take their place. Enki, their chief scientist and Ninhursag their chief medical officer, after getting no satisfactory results splicing animal and Homo Erectus genes, merged their Anunnaki genes with that of Homo Erectus and produced us, Homo Sapiens, a genetically bicameral species, for their purposes as slaves. Because we were a hybrid we could not procreate. The demand for us as workers became greater and we were genetically manipulated to reproduce.

Eventually, we became so numerous that some of us were expelled from the Anunnaki city centers, gradually spreading over the planet. Having become a stable genetic stock and developing more precociously than, perhaps, the Anunnaki had anticipated, the Anunnaki began to be attracted to humans as sexual partners and children were born of these unions. This was unacceptable to the majority of the Anunnaki high council and it was decided to wipe out the human population through a flood that was predictable when Nibiru, the tenth in our solar system and the Anunnaki home planet, came through the inner solar system again (around 12,500 years ago) on one of its periodic 3600 year returns. Some humans were saved by the action of the Anunnaki official, Enki, who was sympathetic to the humans he had originally genetically created. For thousands of years we were their slaves, their workers, their servants, their soldiers in their political battles among themselves.The Anunnaki used us in the construction of their palaces (we retroproject the religious notion of temple on these now), their cities, their mining and refining complexes and their astronomical installations on all the continents. They expanded from Mesopotamia to Egypt to India to South and Central America and the stamp of their presence can be found in the farthest reaches of the planet.

Around 6000 years ago they, probably realizing that they were going to phase off the planet, began to gradually bring humans to independence. Sumer, a human civilization, amazing in its "sudden" and mature and highly advanced character was set up under their tutelage in Mesopotamia. Human kings were inaugurated as go-betweens, foremen of the human populations answering to the Anunnaki. A strain of humans, genetically enhanced with more Anunnaki genes, a bloodline of rulers in a tradition of "servants of the people" was initiated (Gardner). These designated humans were taught technology, mathematics, astronomy, advanced crafts and the ways of advanced civilized society (in schools, called now "mystery schools" but there was no mystery about them). Gardner has brought to light the fact that there exists a robust, highly documented, genealogical, genetic history carrying all the way back to the Anunnaki, possessed by the heterodox tradition of Christianity, which is only now coming forward, no longer gun-shy of the Inquisition. This tradition, preserving the bloodline, is the one branded "heretical" and murderously persecuted by the Roman Church. There were no Dark Ages for this tradition, only for those whom the Church wanted to keep in the dark about the real nature of human history and destroy the bloodline, a direct threat to the power of the Bishops.

The Anunnaki became somewhat more remote from humans. By around 1250 B.C. they had gone into their final phase-out mode. The human population and the foremen kings, now left on their own began to fend for themselves. For some three thousand years, subsequently, we humans have been going through a traumatic transition to racial independence. Proprietary claims made by various groups of humans as to who knew what we should be doing to get the Anunnaki to return or when they returned, perpetuated the palace and social rituals learned under the Anunnaki and sometimes disagreement and strife broke out between them. Religion, as we know it, took form, focused on the "god" or "gods", clearly and unambiguously known to the humans who were in contact with them as imperfect, flesh and blood humanoids, now absent. It was only much later that the Anunnaki were eventually sublimated into cosmic character and status and, later on, conveniently mythologized. We have been dysfunctionally looking to the sky where they went, waiting for Daddy to return to make everything right and tell us what to do for some three thousand years caught in cargo cult religions.

***What Evidence Supports the Sitchin Thesis?***

*The Astronomical Evidence*

A key underpinning of the Sitchin paradigm is the existence, now or in the past, of the tenth planet in our solar system, the home planet of the Anunnaki with the size, orbit, and characteristics described, as Sitchin has demonstrated, in the Enuma Elish and corroborated by Harrington, former chief of the U.S. Naval Observatory, now deceased. Tombaugh discovered Pluto in 1930. Christie, of the U.S. Naval Observatory, discovered Charon, Pluto's moon, in 1978. The characteristics of Pluto derivable from the nature of Charon demonstrated that there must still be a large planet undiscovered because Pluto could not be the cause of the residuals, the "wobbles" in the orbital paths of Uranus and Neptune clearly identifiable. The IRAS (Infrared Astronomical Satellite), during '83 -'84, produced observations of a tenth planet so robust that one of the astronomers on the project said that "all that remains is to name it" — from which point the information has become curiously guarded. In 1992 Harrington and Van Flandern of the Naval Observatory, working with all the information they had at hand, published their findings and opinion that there is, indeed, a tenth planet, even calling it an "intruder" planet. The search was narrowed to the southern skies, below the ecliptic. Harrington invited Sitchin, having read his book and translations of the Enuma Elish, to a meeting at his office and they correlated the current findings with the ancient records.

The recovered Enuma Elish document, a history of the formation of our solar system and more, says that, at the time when Mercury, Venus, Mars, Jupiter, Uranus and Saturn were in place, there was a Uranus sized planet, called Tiamat, in orbit between Mars and Jupiter. Earth was not in place yet. A large wandering planet, called Nibiru, was captured into the system gravitationally. As it passed by the outer planets it caused the anomalies of their moons, the tilting of Uranus on its side, the dislodging of Pluto from its being a moon of Saturn to its own planetary orbit. Its path bent by the gravitational pull of the large planets, first its satellites collided with the large planet Tiamat and, on a second orbit through, Nibiru collided with Tiamat, driving the larger part of it into what is now Earth's orbit to recongeal as Earth, dragging its moon with it to become our Moon with all its anomalies. The shattered debris of Tiamat's smaller part became the asteroid belt, comets, and meteorites. The gouge of our Pacific basin is awesome testimony to the collisional event. Nibiru settled into a 3600 year elliptical retrograde (opposite direction to all the other planets) orbit around our sun, coming in through the asteroid belt region between Mars and Jupiter at perigee and swinging far out past Pluto at apogee. Harrington acknowledged that his information agreed with all these details and the maps they each had drawn of the orbits were almost indistinguishable. The current probable location of Nibiru (Planet X, our tenth) estimated by both was the same.

*The Technological Evidence:* Ooparts is the term used to describe the purportedly out of place in time artifacts, toys, tools, technical devices, depictions and documents which have come to light through archaeological excavation or discovery. Almost everyone is familiar, through published works or documentaries, with the clay pot batteries still containing the electrodes from the Iraqi desert dated at 2500 B.C., the flyable model airplane from a pyramid tomb, the sophisticated machining of stone requiring the most advanced techniques we know today, the 1000 ton precision cut blocks of stone in a temple foundation that we could not even handle, an ancient relief frieze from an Abydos temple depicting rockets, airplanes and even a helicopter, etc. The most recent and quite amazing oopart is the rediscovery of monoatomic gold by David Hudson (Monoatomics are superconductors at room temperature, have anti-gravitic properties and are only now being investigated by the advanced physics community) Hudson's discovery, correlated with the bringing to light, by Gardner, of the suppressed discovery of the Anunnaki gold processing plant on Mt. Horeb by Sir Flinders Petrie in 1889 demonstrates that the monoatomics were already known at least 3000 years ago. These ooparts coupled with evidence from many disciplines and the historical records indicate that an advanced civilization existed in those times possessing a high technology and that that civilization was indeed the Anunnaki.

*The Documentary Evidence:* The recorded historical documentation for the existence and deeds of the Anunnaki has become gradually available to us only since the early 1800's. The excavation of the ancient sites of Mesopotamia brought to light the amazingly advanced civilization of Sumer and, with it, thousands of clay tablets containing not only mundane records of commerce, marriages, military actions and advanced astronomical calculation systems but of the history of the Anunnaki themselves. It is clear from these records that the Sumerians knew these aliens to be real flesh and blood. The library of the ruler, Ashurbanipal, at Nineveh was discovered to have burnt down and the clay tablets held there were fired, preserving them for our reading. One of the most impressive finds, in very recent time, has been a sealed, nine foot by six foot room in Sippar holding, neatly arranged on shelves, a set of some 400 elaborate clay tablets containing an unbroken record of the history of those ancient times, a sort of time capsule. The evidence is so overwhelming and robust that, if it weren't for those with power enough to suppress, it would have been accepted and our world view changed a century ago or, perhaps, sooner.

*The Genetic Evidence:* The recovered records place the location of the Anunnaki laboratory where the first humans were literally produced in east central Africa just above their gold mines. This falls precisely on the map where the mitochondrial DNA "search for Eve" places the first woman Homo-Sapiens and in the same time frame. (The gold mining engineers of Africa have found 100,000 year old gold mines in that area.) The evidence for, and description of advanced genetic engineering is all there in the ancient documents. Our rapid progress from inception to going to Mars soon, after only 250,000 years, does not correspond to the million year periodicities of slow evolutionary development of other species such as Homo Erectus before us. As so many thinkers have pointed out, we are radically and anomalously different, as discussed in this chapter.

*Two Definitive Protocols for Proving or Disproving the*

*Sitchin Paradigm:* A straightforward approach to prove or disprove the Sitchin paradigm is available from astronomy. A thorough, professional search of our solar system should be able, with current technology, to determine the existence or non-existence of a tenth planet with the characteristics of size, orbit, orbital periodicity, and declination from the ecliptic, as Sitchin has determined from the translations of the ancient records and particularly the Enuma Elish document. If that planet is not in our solar system or no evidence can be found for an ejection or cataclysmic destructive event then the Sitchin paradigm falls. This search should be undertaken with highest priority. The matter is sufficiently important to clearing up our historical situation that the academic world should be involved as well as the scientific community. It is the opinion of this author and others that, in light of the evidence already obtained through the use of the Pioneer 10 and 11 and two Voyager space craft, the Infrared Imaging Satellite (IRAS, '83-84) and the clear and unequivocal statements of Harrington when consulting with Sitchin, that the search has already been accomplished, in fact that the planet has already been found.

A second mode of proof or disproof is available through genetics. We will shortly have the entire human genome read out and its details sorted. That process, and the information available as a result, should afford us the opportunity to examine the entire genome for the evidence of the merging of the two species' gene codes, Homo Erectus and Anunnaki, a definitive way to prove or disprove Sitchin's thesis. It is in this process, if proven, that the vital clues to the interpretation of the genome lie as examined in Part 4. It is the position of this author that, not only will the bicameral nature of the code eventually be recognized but valuable clues found through the new paradigm to a profound understanding of the genome and the human.

*The Ramifications of the Sitchin Scenario*

*The First Factor: Background of Conditioning to Subservience:* The overwhelming evidence from the Sitchin paradigm demonstrates conclusively that we were genetically engineered by the Anunnaki and treated as slaves and then limited partners but always as subservient to them. The major ramification (Neil Freer: Breaking the Godspell) of the Sitchin explication is that "religion", as we have known it, is the transmutation, sublimation of the Anunnaki-human relationship of master-subject servitude, of slavery and then limited, subservient partnership. Since then, we have been going through phases of a traumatic transition to species maturity and independence. The major characteristic of this process has been the transmutation in all cultures of them, in particular Enlil/Jehovah in Western culture, into a cosmic deity, commonly called "God" in the Judeo-Christian tradition. (Is this atheism? No, not as such. It simply is a long overdue correction of some local, intra-solar system politics, relatively rather pedestrian in cosmic perspective. Garden variety atheism can now be understood as an early sign of precocious species adolescent rebellion and questioning of the authority of the obviously all too humanoid characteristics of the particular local Nefilim "god" of the Hebrew tribe and Christianity, Yahweh/Jehovah. The new paradigm, once the ancient, subservient godspell is dispelled, simply frees us to go one on one with the universe as our own evolutionary artists and to seek directly whatever unthinkable or thinkable ultimate principle is behind it.) In a later phase we have mythologized them into unreal beings and then, more "sophisticatedly" into psychological archetypes. Only with cumulative evidence, our own evolved technology and restored history through modern scholarship have we now been enabled to grasp the true nature of our genetic creation, our godspell conditioning, traumatic transition, and the opportunity to emerge from species adolescence and amnesia into species maturity. E.O. Wilson, the father of Sociobiology, while rejecting religion as mythology, also sees it as hardwired in the brain as a result of millennia of genetic evolutionary programming. His opinion takes on a whole new meaning in the context of the new paradigm: his description of this predisposition of the human neurological system is an excellent definition of the genetically engineered human displaying the godspell mentality.

*The Second Factor: The Babel Factor, The Splintering of the Human*

*Race Into Hostile Religions:* So, down to our day, incredibly, we have remained still Babel-factored for good crowd control, broken into tribes each proprietarily telling the other that ours is the only accurate tradition of what some particular "god" intended, what rules to follow, what we should be doing to demonstrate we are still loyal and docile servants. Sometimes we just kill each other over it. And persecutions, Crusades, Jihads, Inquisitions, evil empires, the saved and the damned, the martyr, the infidel, the saint, the Protestant, the fundamentalist, the atheist, became — and remain.

*The Third Factor: A Tradition of Absolutism, Suppression and Repression:* Modern scholarship, working on foundations laid as far back as the German school, has reassessed the Old Testament. When the Mesopotamian records began to be unearthed in the 1800's, it was realized that the history revealed from 4000 B.C. forward, was the source of the history of the Old Testament. The Jews, allowed to be confined to captivity by the Babylonians by their "god" Jehovah, had drawn on that history and committed forgery (Sitchin) by rewriting it to their need to establish Enlil/

Jehovah as their supreme ruler. That history shows clearly that two Anunnaki brothers, Enlil and Enki had always been at odds with each other personally and politically. Enki was the Anunnaki scientist who, together with his half-sister scientist, Ninhursag, had created humans and had always been favorable to them. Enlil had always had reservations about humans and dominated them with severity.

Gardner has made the case that when the Old Testament referred to The Lord (Adonai, Enki) it was when good things were being done to or for the Hebrews. When terrible things were being done or allowed to happen to them, the reference was always to Jehovah(Yahweh, Enlil) and it was to Enlil's dominance that they finally capitulated and whom they worshipped in fear. The bloodline of human leaders, further enhanced with Anunnaki genes, initiated by Enki (The Lord, Adonai), passed from Sumer through Egypt to Israel through David and the messiahs (anointed ones), fostered by the Essene communities as revealed in the Dead Sea Scrolls. It was recognized but treated ambiguously by the large faction of the Hebrews who paid allegiance to Enlil.

The strain of Hellenistic, liberal, Judaism (Jesus was in the bloodline) eventually became known as heterodox Christian was opposed by the Pauline branch which eventually allied itself with Rome and became the Roman Church. The orthodox Roman Church perpetuated the Enlil type fear and subservience tradition and, in its turn, suppressed, persecuted and brutalized the human-centered strain of the bloodline.. The reason why women have always been denigrated and suppressed by the Roman Church is genetic: the bloodline was dependent on transmission through the women of the line and any recognition of them would be recognition of the line.

It is manifestly unjust that the geneticist is deprived of this vital block of information about the genome due to the preclusive attitude of the "bishops" still reigning in archaeology. The restoration of our true human history clearly affords resolution to the perennial Creationist-Evolutionary as well as other conflicts and will lead to a consensual, generic definition of a human and a planetary unity never before experienced. By that release, it also reveals clues to our current genetic questions. These are best approached through the answers to the most obvious objections that have been raised to the new paradigm.

*Scientific Objections to the Thesis:* How could the Anunnaki, clearly described as comfortable in earth gravity and atmosphere, very similar to current humans in all ways, have evolved on a planet within our solar system whose orbital apogee takes it into the deep cold of space for much of its orbit? The ancient records repeatedly describe Nibiru as a "radiant" planet. This may be understood as having a high core temperature. Although controversial, there is also astrophysical opinion that a large body in elongated orbit is constantly tending toward circular orbit and this causes stresses in the body that could generate a good deal of heat. That their planet is gradually cooling, may be indicated by Sitchin's interpretation of their colonizing Earth (contains most of the gold identifiable in the solar system) for the purpose of obtaining large quantities of gold for molecular seeding of their atmosphere with a reflective gold shielding. Pertinent here is Harrington's confident statement to Sitchin that it is "a nice, good planet, could be surrounded by gases, probably has an atmosphere and could support life like ours". The sunlight level there might be quite different than on earth. Although not always, the Anunnaki were often depicted or sculpted with what seem to be obviously sunglasses.

If, however, the Anunnaki evolved on a radically different planet from earth under quite different conditions to which to adapt, why should they have turned out to be so identical to to human species? Sitchin's answer is based on the collisional event between the intruder planet, Nibiru, and the planet Tiamat, the residual part of which recongealed into the Earth after being driven into current Earth orbit. That the two, or at least one, of the colliding planets was sufficiently developed to have evolved basic organic compounds, perhaps even simple life, the cross-seeding of everything from amino acids to more complicated organic compounds or even primitive organisms, could account for the evolutionary similarity. Although this author finds it a reasonable hypothesis, even trivial, that advanced civilizations would be capable of crossing extremely different genomes, perhaps with even radically different bases, the cross-seeding theory can account for the apparent relative ease with which the Anunnaki impinged their genes on the genes of Homo Erectus. The Anunnaki skill level, 200,000 years ago, is indicated well by the recorded fact that, in early trials, they succeeded in crossing animal genes with Homo Erectus genes, obtained living hybrids but never a satisfactory product which led them to modify Homo Erectus genes with their advanced genes.

This provides a comprehensive context in which to understand and explain the enigmatic facts of our being contemporaneous with Neanderthals or even preceding them and other anthropological anomalies — at least within the Darwinian model — of our species existence. We have 46 chromosomes, primates have 48 and the fusion of the second and third chromosomes in primates is a mystery. Even today the anthropological sector is scrambling to find a viable ancestor species for us, Homo Erectus is currently being promoted; we showed up too suddenly in the chronology of the fossil record; our 4000 plus genetic diseases are a total enigma.

We present with very many startling, obvious differences from primates, and those differences, suddenly appearing in our species, are radical: we have foreheads, hardly any brow ridges, eye sockets far more rectangular than round; relatively tiny nasal passages; small flat mouths and a chin; far less muscular strength and bone density; our skin, sweat process and glands, body hair, throats, and salt management are completely different. Human females don't have an estrus cycle. We are bipedal. Our brains are remarkably different to say the very least.

If our genome is estimated as 95% to 98% similar to the chimpanzee, how could there be a melding of the Homo Erectus and Anunnaki genomes, or impingement of the advanced code on the lesser advanced one detectable? The author suggests that this is a major question probably answerable only by the geneticists open-minded enough to attack it. The resolution of that question should provide rich additional clues in itself.

### *Significant Genetic Ramifications, Clues and Indicators Afforded by the Sitchin Thesis and its Restoration of our True History*

If it is taken as archaeologically and historically demonstrated that we are a genetically engineered, bicameral species, the product of genetically melding two racial genomes, the past becomes a rich archive of anthropological, technical, historical and especially genetic information and data. Just as the reluctant acknowledgement that ancient records from the Sumerian or Chinese civilizations contained accurate astronomical observations and data opened up a valuable resource to modern astronomers, so the history of our species' genetic creation in a laboratory pinpointed on the map of east central Africa precisely where the mitochondrial "search for Eve" locates the first human female(s), opens an astounding resource to modern geneticists.

The most important advantage is that, enabled with this knowledge and perspective, geneticists can begin to understand and decipher and interpret our genome with the tremendous advantage of working within a robust context providing major clues as to what to look for and where to look and why.

The ancient records of why, when, and how we were genetically engineered make it abundantly clear that we were brought into existence as a subordinate species, a slave species to relieve the Anunnaki miner echelons. (The essential, detailed documentation through translations and illustrations of the actual genetic processes (in vitro; cloning, etc. ) used by the Anunnaki is found in Sitchin, The Twelfth Planet, chapter 12) It is specified pointedly that, although the Anunnaki lived, literally, extended lifetimes of thousands of our years (either because of the way they themselves had evolved on their home planet or, perhaps, because of their genetic engineering capabilities to achieve that longevity, and possibly through their use of the monoatomic form of gold) they deliberately did not bestow that potential on us. In fact, it is mentioned clearly that they deliberately withheld it. This deliberate withholding of immortality and, perhaps, even a shortening of longevity, may provide a major clue to our aging process and mortality. (Sociologically, this is the reason why immortality, in one form or another, has become the sublimated reward for a good life in all the religions.). From the details given in the ancient records, it is conceivable that some engineering of the process was executed deliberately to suppress certain characteristics to make better and docile slaves.

If they were interested only in engineering what clearly amounted to disposable units then it may reasonably be inferred that the quality and completeness of the engineering would not have to have been taken to the maximum. This is, in the opinion of this author, the basis for the four thousand plus genetic diseases and defects we present. It has been argued theoretically that, although the vast majority of the species on this planet present with only a few typical genetic diseases, we show 4000 because of the relative complexity of our organism. But we are not that much more complex genetically than even the higher primates and this argument does not hold.

Several significant details in the records of our genetic genesis may hold clues. It is recounted that two kinds of females were created, those who would bear children and those who would not. Determining genetically how we, as sterile mutants when first created, were manipulated to be able to procreate may be a major lead. Finally, knowing that an advanced bloodline of humans, enhanced by additional Anunnaki genes, was created around 4000 B.C. is even more valuable a clue since that bloodline has been carefully nurtured and protected through to our day and, therefore, available for investigation and analysis and comparison.

A primary practical tool would be the devising of a protocol through which a crossing of the two gene codes might be recognized. The ancient records could be interpreted to mean either a complete melding of the two codes or, alternatively, the impingement of selected Anunnaki genes on the Homo Erectus code to tweak up the more primitive code to at least a condition of intelligence and physical competence to handle the mining of gold. It is possible that in some 200,000 years of our existence the accommodation between the codes has smoothed and recognition may be difficult indeed. The obtaining of some robust Homo Erectus

DNA samples from fossils would be a great help. It may evolve that, in working on the genetic diseases from the perspective of the genome being bicameral, those defects may yield an indirect key in some pattern or mechanism that indicates the nature and extent of the splicing of the codes.

During the genetic investigation now in process, therefore, it would be valuable indeed for geneticists, at the minimum, to be constantly inspecting the results of the decipherment for signs of the genetic merging, and to develop protocols to determine such, if needed, as well as following the clues mentioned herein. By doing so, interpretations and explanations may possibly be facilitated, progress accelerated and a far more comprehensive overview of the genome achieved. The information gained would, reciprocally, be a major, pivotal, invaluable resultant spin-off contribution to our species' evolutionary, anthropological, and cultural generic history.Introduction to Evolutionary Biology. Evolution is the cornerstone of modern biology. It unites all the fields of biology under one theoretical umbrella. It is not a difficult concept, but very few people — the majority of biologists included — have a satisfactory grasp of it. One common mistake is believing that species can be arranged on an evolutionary ladder from bacteria through "lower" animals, to "higher" animals and, finally, up to man. Mistakes permeate popular science expositions of evolutionary biology. Mistakes even filter into biology journals and texts. For example, Lodish, et. al., in their cell biology text, proclaim, "It was Charles Darwin's great insight that organisms are all related in a great chain of being..." In fact, the idea of a great chain of being, which traces to Linnaeus, was overturned by Darwin's idea of common descent.

Misunderstandings about evolution are damaging to the study of evolution and biology as a whole. People who have a general interest in science are likely to dismiss evolution as a soft science after absorbing the pop science nonsense that abounds. The impression of it being a soft science is reinforced when biologists in unrelated fields speculate publicly about evolution.

This is a brief introduction to evolutionary biology. I attempt to explain basics of the theory of evolution and correct many of the misconceptions.

*What is Evolution?:* Evolution is a change in the gene pool of a population over time. A gene is a hereditary unit that can be passed on unaltered for many generations. The gene pool is the set of all genes in a species or population.

The English moth, *Biston betularia*, is a frequently cited example of observed evolution. [evolution: a change in the gene pool] In this moth there are two color morphs, light and dark. H. B. D. Kettlewell found that dark moths constituted less than 2% of the population prior to 1848. The frequency of the dark morph increased in the years following. By 1898, the 95% of the moths in Manchester and other highly industrialized areas were of the dark type. Their frequency was less in rural areas. The moth population changed from mostly light colored moths to mostly dark colored moths. The moths' color was primarily determined by a single gene. [gene: a hereditary unit] So, the change in frequency of dark colored moths represented a change in the gene pool. [gene pool: the set all of genes in a population] This change was, by definition, evolution.

The increase in relative abundance of the dark type was due to natural selection. The late eighteen hundreds was the time of England's industrial revolution. Soot from factories darkened the birch trees the moths landed on. Against a sooty background, birds could see the lighter colored moths better and ate more of them. As a result, more dark moths survived until reproductive age and left offspring. The greater number of offspring left by dark moths is what caused their increase in frequency. This is an example of natural selection.

Populations evolve. [evolution: a change in the gene pool] In order to understand evolution, it is necessary to view populations as a collection of individuals, each harboring a different set of traits. A single organism is never typical of an entire population unless there is no variation within that population. Individual organisms do not evolve, they retain the same genes throughout their life. When a population is evolving, the ratio of different genetic types is changing — each individual organism within a population does not change. For example, in the previous example, the frequency of black moths increased; the moths did not turn from light to gray to dark in concert. The process of evolution can be summarized in three sentences: Genes mutate. [gene: a hereditary unit] Individuals are selected. Populations evolve.

Evolution can be divided into microevolution and macroevolution. The kind of evolution documented above is microevolution. Larger changes, such as when a new species is formed, are called macroevolution. Some biologists feel the mechanisms of macroevolution are different from those of microevolutionary change. Others think the distinction between the two is arbitrary — macroevolution is cumulative microevolution.

The word evolution has a variety of meanings. The fact that all organisms are linked via descent to a common ancestor is often called evolution. The theory of how the first living organisms appeared is often called evolution. This should be called abiogenesis. And frequently, people use the word evolution when they really mean natural selection — one of the many mechanisms of evolution.

*Common Misconceptions about Evolution:* Evolution can occur without morphological change; and morphological change can occur without evolution. Humans are larger now than in the recent past, a result of better diet and medicine. Phenotypic changes, like this, induced solely by changes in environment do not count as evolution because they are not heritable; in other words the change is not passed on to the organism's offspring. Phenotype is the morphological, physiological, biochemical, behavioral and other properties exhibited by a living organism. An organism's phenotype is determined by its genes and its environment. Most changes due to environment are fairly subtle, for example size differences. Large scale phenotypic changes are obviously due to genetic changes, and therefore are evolution.

Evolution is not progress. Populations simply adapt to their current surroundings. They do not necessarily become better in any absolute sense over time. A trait or strategy that is successful at one time may be unsuccessful at another. Paquin and Adams demonstrated this experimentally. They founded a yeast culture and maintained it for many generations. Occasionally, a mutation would arise that allowed its bearer to reproduce better than its contemporaries. These mutant strains would crowd out the formerly dominant strains. Samples of the most successful strains from the culture were taken at a variety of times. In later competition experiments, each strain would outcompete the immediately previously dominant type in a culture. However, some earlier isolates could outcompete strains that arose late in the experiment. Competitive ability of a strain was always better than its previous type, but competitiveness in a general sense was not increasing. Any organism's success depends on the behavior of its contemporaries. For most traits or behaviors there is likely no optimal design or strategy, only contingent ones. Evolution can be like a game of paper/scissors/rock.

Organisms are not passive targets of their environment. Each species modifies its own environment. At the least, organisms remove nutrients from and add waste to their surroundings. Often, waste products benefit other species. Animal dung is fertilizer for plants. Conversely, the oxygen we breathe is a waste product of plants. Species do not simply change to fit their environment; they modify their environment to suit them as well. Beavers build a dam to create a pond suitable to sustain them and raise young. Alternately, when the environment changes, species can migrate to suitable climes or seek out microenvironments to which they are adapted.

*Genetic Variation:* Evolution requires genetic variation. If there were no dark moths, the population could not have evolved from mostly light to mostly dark. In order for continuing evolution there must be mechanisms to increase or create genetic variation and mechanisms to decrease it. Mutation is a change in a gene. These changes are the source of new genetic variation. Natural selection operates on this variation.

Genetic variation has two components: allelic diversity and non- random associations of alleles. Alleles are different versions of the same gene. For example, humans can have A, B or O alleles that determine one aspect of their blood type. Most animals, including humans, are diploid — they contain two alleles for every gene at every locus, one inherited from their mother and one inherited from their father. Locus is the location of a gene on a chromosome. Humans can be AA, AB, AO, BB, BO or OO at the blood group locus. If the two alleles at a locus are the same type (for instance two A alleles) the individual would be called homozygous. An individual with two different alleles at a locus (for example, an AB individual) is called heterozygous. At any locus there can be many different alleles in a population, more alleles than any single organism can possess. For example, no single human can have an A, B and an O allele.

Considerable variation is present in natural populations. At 45 percent of loci in plants there is more than one allele in the gene pool. [allele: alternate version of a gene (created by mutation)] Any given plant is likely to be heterozygous at about 15 percent of its loci. Levels of genetic variation in animals range from roughly 15% of loci having more than one allele (polymorphic) in birds, to over 50% of loci being polymorphic in insects. Mammals and reptiles are polymorphic at about 20% of their loci - - amphibians and fish are polymorphic at around 30% of their loci. In most populations, there are enough loci and enough different alleles that every individual, identical twins excepted, has a unique combination of alleles.

Linkage disequilibrium is a measure of association between alleles of two different genes. [allele: alternate version of a gene] If two alleles were found together in organisms more often than would be expected, the alleles are in linkage disequilibrium. If there two loci in an organism (A and B) and two alleles at each of these loci (A1, A2, B1 and B2) linkage disequilibrium (D) is calculated as D = f(A1B1) * f(A2B2) - f(A1B2) * f(A2B1) (where f(X) is the frequency of X in the population). [Loci (plural of locus): location of a gene on a chromosome] D varies between -1/4 and 1/4; the greater the deviation from zero, the greater the linkage. The sign is simply a consequence of how the alleles are numbered. Linkage disequilibrium can be the result of physical proximity of the genes. Or, it can be maintained by natural selection if some combinations of alleles work better as a team.

Natural selection maintains the linkage disequilibrium between color and pattern alleles in *Papilio memnon*. [linkage disequilibrium: association between alleles at different loci] In this moth species, there is a gene that determines wing morphology. One allele at this locus leads to a moth that has a tail; the other allele codes for a untailed moth. There is another gene that determines if the wing is brightly or darkly colored. There are thus four possible types of moths: brightly colored moths with and without tails, and dark moths with and without tails. All four can be produced when moths are brought into the lab and bred. However, only two of these types of moths are found in the wild: brightly colored moths with tails and darkly colored moths without tails. The non-random association is maintained by natural selection. Bright, tailed moths mimic the pattern of an unpalatable species. The dark morph is cryptic. The other two combinations are neither mimetic nor cryptic and are quickly eaten by birds.

Assortative mating causes a non-random distribution of alleles at a single locus. [locus: location of a gene on a chromosome] If there are two alleles (A and a) at a locus with frequencies p and q, the frequency of the three possible genotypes (AA, Aa and aa) will be $p^2$, 2pq and $q^2$, respectively. For example, if the frequency of A is 0.9 and the frequency of a is 0.1, the frequencies of AA, Aa and aa individuals are: 0.81, 0.18 and 0.01. This distribution is called the Hardy-Weinberg equilibrium.

Non-random mating results in a deviation from the Hardy-Weinberg distribution. Humans mate assortatively according to race; we are more likely to mate with someone of own race than another. In populations that mate this way, fewer heterozygotes are found than would be predicted under random mating. [heterozygote: an organism that has two different alleles at a locus] A decrease in heterozygotes can be the result of mate choice, or simply the result of population subdivision. Most organisms have a limited dispersal capability, so their mate will be chosen from the local population.

*Evolution within a Lineage* : In order for continuing evolution there must be mechanisms to increase or create genetic variation and mechanisms to decrease it. The mechanisms of evolution are mutation, natural selection, genetic drift, recombination and gene flow. I have grouped them into two classes — those that decrease genetic variation and those that increase it.

### *Mechanisms that Decrease Genetic Variation*

#### *Natural Selection*

Some types of organisms within a population leave more offspring than others. Over time, the frequency of the more prolific type will increase. The difference in reproductive capability is called natural selection. Natural selection is the only mechanism of adaptive evolution; it is defined as differential reproductive success of pre- existing classes of genetic variants in the gene pool.

The most common action of natural selection is to remove unfit variants as they arise via mutation. [natural selection: differential reproductive success of genotypes] In other words, natural selection usually prevents new alleles from increasing in frequency. This led a famous evolutionist, George Williams, to say "Evolution proceeds in spite of natural selection."

Natural selection can maintain or deplete genetic variation depending on how it acts. When selection acts to weed out deleterious alleles, or causes an allele to sweep to fixation, it depletes genetic variation. When heterozygotes are more fit than either of the homozygotes, however, selection causes genetic variation to be maintained. [heterozygote: an organism that has two different alleles at a locus. | homozygote: an organism that has two identical alleles at a locus] This is called balancing selection. An example of this is the maintenance of sickle-cell alleles in human populations subject to malaria. Variation at a single locus determines whether red blood cells are shaped normally or sickled. If a human has two alleles for sickle-cell, he/she develops anemia — the shape of sickle-cells precludes them carrying normal levels of oxygen. However, heterozygotes who have one copy of the sickle-cell allele, coupled with one normal allele enjoy some resistance to malaria — the shape of sickled cells make it harder for the plasmodia (malaria causing agents) to enter the cell. Thus, individuals homozygous for the normal allele suffer more malaria than heterozygotes. Individuals homozygous for the sickle- cell are anemic. Heterozygotes have the highest fitness of these three types. Heterozygotes pass on both sickle-cell and normal alleles to the next generation. Thus, neither allele can be eliminated from the gene pool. The sickle-cell allele is at its highest frequency in regions of Africa where malaria is most pervasive.

Balancing selection is rare in natural populations. [balancing selection: selection favoring heterozygotes] Only a handful of other cases beside the sickle-cell example have been found. At one time population geneticists thought balancing selection could be a general explanation for the levels of genetic variation found in natural populations. That is no longer the case. Balancing selection is only rarely found in natural populations. And, there are theoretical reasons why natural selection cannot maintain polymorphisms at several loci via balancing selection.

Individuals are selected. The example I gave earlier was an example of evolution via natural selection. [natural selection: differential reproductive success of genotypes] Dark colored moths had a higher reproductive success because light colored moths suffered a higher predation rate. The decline of light colored alleles was caused by light colored individuals being removed from the gene pool (selected against). Individual organisms either reproduce or fail to reproduce and are hence the unit of selection. One way alleles can change in frequency is to be housed in organisms with different reproductive rates. Genes are not the unit of selection (because their success depends on the organism's other genes as well); neither are groups of organisms a unit of selection. There are some exceptions to this "rule," but it is a good generalization.

Organisms do not perform any behaviors that are for the good of their species. An individual organism competes primarily with others of it own species for its reproductive success. Natural selection favors selfish behavior because any truly altruistic act increases the recipient's reproductive success while lowering the donors. Altruists would disappear from a population as the non- altruists would reap the benefits, but not pay the costs, of altruistic acts. Many behaviors appear altruistic. Biologists, however, can demonstrate that these behaviors are only apparently altruistic. Cooperating with or helping other organisms is often the most selfish strategy for an animal. This is called reciprocal altruism. A good example of this is blood sharing in vampire bats. In these bats, those lucky enough to find a meal will often share part of it with an unsuccessful bat by regurgitating some blood into the other's mouth. Biologists have found that these bats form bonds with partners and help each other out when the other is needy. If a bat is found to be a "cheater," (he accepts blood when starving, but does not donate when his partner is) his partner will abandon him. The bats are thus not helping each other altruistically; they form pacts that are mutually beneficial.

Helping closely related organisms can appear altruistic; but this is also a selfish behavior. Reproductive success (fitness) has two components; direct fitness and indirect fitness. Direct fitness is a measure of how many alleles, on average, a genotype contributes to the subsequent generation's gene pool by reproducing. Indirect fitness is a measure of how many alleles identical to its own it helps to enter the gene pool. Direct fitness plus indirect fitness is inclusive fitness. J. B. S. Haldane once remarked he would gladly drown, if by doing so he saved two siblings or eight cousins. Each of his siblings would share one half his alleles; his cousins, one eighth. They could potentially add as many of his alleles to the gene pool as he could.

Natural selection favors traits or behaviors that increase a genotype's inclusive fitness. Closely related organisms share many of the same alleles. In diploid species, siblings share on average at least 50% of their alleles. The percentage is higher if the parents are related. So, helping close relatives to reproduce gets an organism's own alleles better represented in the gene pool. The benefit of helping relatives increases dramatically in highly inbred species. In some cases, organisms will completely forgo reproducing and only help their relatives reproduce. Ants, and other eusocial insects, have sterile castes that only serve the queen and assist her reproductive efforts. The sterile workers are reproducing by proxy.

The words selfish and altruistic have connotations in everyday use that biologists do not intend. Selfish simply means behaving in such a way that one's own inclusive fitness is maximized; altruistic means behaving in such a way that another's fitness is increased at the expense of ones' own. Use of the words selfish and altruistic is not meant to imply that organisms consciously understand their motives.

The opportunity for natural selection to operate does not induce genetic variation to appear — selection only distinguishes between existing variants. Variation is not possible along every imaginable axis, so all possible adaptive solutions are not open to populations. To pick a somewhat ridiculous example, a steel shelled turtle might be an improvement over regular turtles. Turtles are killed quite a bit by cars these days because when confronted with danger, they retreat into their shells — this is not a great strategy against a two ton automobile. However, there is no variation in metal content of shells, so it would not be possible to select for a steel shelled turtle.

Here is a second example of natural selection. Geospiza fortis lives on the Galapagos islands along with fourteen other finch species. It feeds on the seeds of the plant Tribulus cistoides, specializing on the smaller seeds. Another species, G. Magnirostris, has a larger beak and specializes on the larger seeds. The health of these bird populations depends on seed production. Seed production, in turn, depends on the arrival of wet season. In 1977, there was a drought. Rainfall was well below normal and fewer seeds were produced. As the season progressed, the G. fortis population depleted the supply of small seeds. Eventually, only larger seeds remained. Most of the finches starved; the population plummeted from about twelve hundred birds to less than two hundred. Peter Grant, who had been studying these finches, noted that larger beaked birds fared better than smaller beaked ones. These larger birds had offspring with correspondingly large beaks. Thus, there was an increase in the proportion of large beaked birds in the population the next generation. To prove that the change in bill size in Geospiza fortis was an evolutionary change, Grant had to show that differences in bill size were at least partially genetically based. He did so by crossing finches of various beak sizes and showing that a finch's beak size was influenced by its parent's genes. Large beaked birds had large beaked offspring; beak size was not due to environmental differences (in parental care, for example).

Natural selection may not lead a population to have the optimal set of traits. In any population, there would be a certain combination of possible alleles that would produce the optimal set of traits (the global optimum); but there are other sets of alleles that would yield a population almost as adapted (local optima). Transition from a local optimum to the global optimum may be hindered or forbidden because the population would have to pass through less adaptive states to make the transition. Natural selection only works to bring populations to the nearest optimal point. This idea is Sewall Wright's adaptive landscape. This is one of the most influential models that shape how evolutionary biologists view evolution.

Natural selection does not have any foresight. It only allows organisms to adapt to their current environment. Structures or behaviors do not evolve for future utility. An organism adapts to its environment at each stage of its evolution. As the environment changes, new traits may be selected for. Large changes in populations are the result of cumulative natural selection. Changes are introduced into the population by mutation; the small minority of these changes that result in a greater reproductive output of their bearers are amplified in frequency by selection.

Complex traits must evolve through viable intermediates. For many traits, it initially seems unlikely that intermediates would be viable. What good is half a wing? Half a wing may be no good for flying, but it may be useful in other ways. Feathers are thought to have evolved as insulation (ever worn a down jacket?) and/or as a way to trap insects. Later, proto-birds may have learned to glide when leaping from tree to tree. Eventually, the feathers that originally served as insulation now became co-opted for use in flight. A trait's current utility is not always indicative of its past utility. It can evolve for one purpose, and be used later for another. A trait evolved for its current utility is an adaptation; one that evolved for another utility is an exaptation. An example of an exaptation is a penguin's wing. Penguins evolved from flying ancestors; now they are flightless and use their wings for swimming.

*Common Misconceptions about Selection*

Selection is not a force in the sense that gravity or the strong nuclear force is. However, for the sake of brevity, biologists sometimes refer to it that way. This often leads to some confusion when biologists speak of selection "pressures." This implies that the environment "pushes" a population to more adapted state. This is not the case. Selection merely favors beneficial genetic changes when they occur by chance — it does not contribute to their appearance. The potential for selection to act may long precede the appearance of selectable genetic variation. When selection is spoken of as a force, it often seems that it is has a mind of its own; or as if it was nature personified. This most often occurs when biologists are waxing poetic about selection. This has no place in scientific discussions of evolution. Selection is not a guided or cognizant entity; it is simply an effect.

A related pitfall in discussing selection is anthropomorphizing on behalf of living things. Often conscious motives are seemingly imputed to organisms, or even genes, when discussing evolution. This happens most frequently when discussing animal behavior. Animals are often said to perform some behavior because selection will favor it. This could more accurately worded as "animals that, due to their genetic composition, perform this behavior tend to be favored by natural selection relative to those who, due to their genetic composition, don't." Such wording is cumbersome. To avoid this, biologists often anthropomorphize. This is unfortunate because it often makes evolutionary arguments sound silly. Keep in mind this is only for convenience of expression.

The phrase "survival of the fittest" is often used synonymously with natural selection. The phrase is both incomplete and misleading. For one thing, survival is only one component of selection — and perhaps one of the less important ones in many populations. For example, in polygynous species, a number of males survive to reproductive age, but only a few ever mate. Males may differ little in their ability to survive, but greatly in their ability to attract mates — the difference in reproductive success stems mainly from the latter consideration. Also, the word fit is often confused with physically fit. Fitness, in an evolutionary sense, is the average reproductive output of a class of genetic variants in a gene pool. Fit does not necessarily mean biggest, fastest or strongest.

*Sexual Selection*

In many species, males develop prominent secondary sexual characteristics. A few oft cited examples are the peacock's tail, coloring and patterns in male birds in general, voice calls in frogs and flashes in fireflies. Many of these traits are a liability from the standpoint of survival. Any ostentatious trait or noisy, attention getting behavior will alert predators as well as potential mates. How then could natural selection favor these traits?

Natural selection can be broken down into many components, of which survival is only one. Sexual attractiveness is a very important component of selection, so much so that biologists use the term sexual selection when they talk about this subset of natural selection.

Sexual selection is natural selection operating on factors that contribute to an organism's mating success. Traits that are a liability to survival can evolve when the sexual attractiveness of a trait outweighs the liability incurred for survival. A male who lives a short time, but produces many offspring is much more successful than a long lived one that produces few. The former's genes will eventually dominate the gene pool of his species. In many species, especially polygynous species where only a few males monopolize all the females, sexual selection has caused pronounced sexual dimorphism. In these species males compete against other males for mates. The competition can be either direct or mediated by female choice. In species where females choose, males compete by displaying striking phenotypic characteristics and/or performing elaborate courtship behaviors. The females then mate with the males that most interest them, usually the ones with the most outlandish displays. There are many competing theories as to why females are attracted to these displays.

The good genes model states that the display indicates some component of male fitness. A good genes advocate would say that bright coloring in male birds indicates a lack of parasites. The females are cueing on some signal that is correlated with some other component of viability.

Selection for good genes can be seen in sticklebacks. In these fish, males have red coloration on their sides. Milinski and Bakker showed that intensity of color was correlated to both parasite load and sexual attractiveness. Females preferred redder males. The redness indicated that he was carrying fewer parasites.

Evolution can get stuck in a positive feedback loop. Another model to explain secondary sexual characteristics is called the runaway sexual selection model. R. A. Fisher proposed that females may have an innate preference for some male trait before it appears in a population. Females would then mate with male carriers when the trait appears. The offspring of these matings have the genes for both the trait and the preference for the trait. As a result, the process snowballs until natural selection brings it into check. Suppose that female birds prefer males with longer than average tail feathers. Mutant males with longer than average feathers will produce more offspring than the short feathered males. In the next generation, average tail length will increase. As the generations progress, feather length will increase because females do not prefer a specific length tail, but a longer than average tail. Eventually tail length will increase to the point were the liability to survival is matched by the sexual attractiveness of the trait and an equilibrium will be established. Note that in many exotic birds male plumage is often very showy and many species do in fact have males with greatly elongated feathers. In some cases these feathers are shed after the breeding season.

None of the above models are mutually exclusive. There are millions of sexually dimorphic species on this planet and the forms of sexual selection probably vary amongst them.

### *Genetic Drift*

Allele frequencies can change due to chance alone. This is called genetic drift. Drift is a binomial sampling error of the gene pool. What this means is, the alleles that form the next generation's gene pool are a sample of the alleles from the current generation. When sampled from a population, the frequency of alleles differs slightly due to chance alone.

Alleles can increase or decrease in frequency due to drift. The average expected change in allele frequency is zero, since increasing or decreasing in frequency is equally probable. A small percentage of alleles may continually change frequency in a single direction for several generations just as flipping a fair coin may, on occasion, result in a string of heads or tails. A very few new mutant alleles can drift to fixation in this manner.

In small populations, the variance in the rate of change of allele frequencies is greater than in large populations. However, the overall rate of genetic drift (measured in substitutions per generation) is independent of population size. [genetic drift: a random change in allele frequencies] If the mutation rate is constant, large and small populations lose alleles to drift at the same rate. This is because large populations will have more alleles in the gene pool, but they will lose them more slowly. Smaller populations will have fewer alleles, but these will quickly cycle through. This assumes that mutation is constantly adding new alleles to the gene pool and selection is not operating on any of these alleles.

Sharp drops in population size can change allele frequencies substantially. When a population crashes, the alleles in the surviving sample may not be representative of the precrash gene pool. This change in the gene pool is called the founder effect, because small populations of organisms that invade a new territory (founders) are subject to this. Many biologists feel the genetic changes brought about by founder effects may contribute to isolated populations developing reproductive isolation from their parent populations. In sufficiently small populations, genetic drift can counteract selection. [genetic drift: a random change in allele frequencies] Mildly deleterious alleles may drift to fixation.

Wright and Fisher disagreed on the importance of drift. Fisher thought populations were sufficiently large that drift could be neglected. Wright argued that populations were often divided into smaller subpopulations. Drift could cause allele frequency differences between subpopulations if gene flow was small enough. If a subpopulation was small enough, the population could even drift through fitness valleys in the adaptive landscape. Then, the subpopulation could climb a larger fitness hill. Gene flow out of this subpopulation could contribute to the population as a whole adapting. This is Wright's Shifting Balance theory of evolution.

Both natural selection and genetic drift decrease genetic variation. If they were the only mechanisms of evolution, populations would eventually become homogeneous and further evolution would be impossible. There are, however, mechanisms that replace variation depleted by selection and drift. These are discussed below.

### *Mechanisms that Increase Genetic Variation*

#### *Mutation*

The cellular machinery that copies DNA sometimes makes mistakes. These mistakes alter the sequence of a gene. This is called a mutation. There are many kinds of mutations. A point mutation is a mutation in which one "letter" of the genetic code is changed to another. Lengths of DNA can also be deleted or inserted in a gene; these are also mutations. Finally, genes or parts of genes can become inverted or duplicated. Typical rates of mutation are between $10^{-10}$ and $10^{-12}$ mutations per base pair of DNA per generation.

Most mutations are thought to be neutral with regards

to fitness. (Kimura defines neutral as $|s| < 1/2Ne$, where s is the selective coefficient and Ne is the effective population size.) Only a small portion of the genome of eukaryotes contains coding segments. And, although some non-coding DNA is involved in gene regulation or other cellular functions, it is probable that most base changes would have no fitness consequence.

Most mutations that have any phenotypic effect are deleterious. Mutations that result in amino acid substitutions can change the shape of a protein, potentially changing or eliminating its function. This can lead to inadequacies in biochemical pathways or interfere with the process of development. Organisms are sufficiently integrated that most random changes will not produce a fitness benefit. Only a very small percentage of mutations are beneficial. The ratio of neutral to deleterious to beneficial mutations is unknown and probably varies with respect to details of the locus in question and environment.

Mutation limits the rate of evolution. The rate of evolution can be expressed in terms of nucleotide substitutions in a lineage per generation. Substitution is the replacement of an allele by another in a population. This is a two step process: First a mutation occurs in an individual, creating a new allele. This allele subsequently increases in frequency to fixation in the population. The rate of evolution is $k = 2Nvu$ (in diploids) where k is nucleotide substitutions, N is the effective population size, v is the rate of mutation and u is the proportion of mutants that eventually fix in the population.

Mutation need not be limiting over short time spans. The rate of evolution expressed above is given as a steady state equation; it assumes the system is at equilibrium. Given the time frames for a single mutant to fix, it is unclear if populations are ever at equilibrium. A change in environment can cause previously neutral alleles to have selective values; in the short term evolution can run on "stored" variation and thus is independent of mutation rate. Other mechanisms can also contribute selectable variation. Recombination creates new combinations of alleles (or new alleles) by joining sequences with separate microevolutionary histories within a population. Gene flow can also supply the gene pool with variants. Of course, the ultimate source of these variants is mutation.

*The Fate of Mutant Alleles*

Mutation creates new alleles. Each new allele enters the gene pool as a single copy amongst many. Most are lost from the gene pool, the organism carrying them fails to reproduce, or reproduces but does not pass on that particular allele. A mutant's fate is shared with the genetic background it appears in. A new allele will initially be linked to other loci in its genetic background, even loci on other chromosomes. If the allele increases in frequency in the population, initially it will be paired with other alleles at that locus — the new allele will primarily be carried in individuals heterozygous for that locus. The chance of it being paired with itself is low until it reaches intermediate frequency. If the allele is recessive, its effect won't be seen in any individual until a homozygote is formed. The eventual fate of the allele depends on whether it is neutral, deleterious or beneficial.

*Neutral alleles*

Most neutral alleles are lost soon after they appear. The average time (in generations) until loss of a neutral allele is 2(Ne/N) ln(2N) where N is the effective population size (the number of individuals contributing to the next generation's gene pool) and N is the total population size. Only a small percentage of alleles fix. Fixation is the process of an allele increasing to a frequency at or near one. The probability of a neutral allele fixing in a population is equal to its frequency. For a new mutant in a diploid population, this frequency is 1/2N.

If mutations are neutral with respect to fitness, the rate of substitution (k) is equal to the rate of mutation(v). This does not mean every new mutant eventually reaches fixation. Alleles are added to the gene pool by mutation at the same rate they are lost to drift. For neutral alleles that do fix, it takes an average of 4N generations to do so. However, at equilibrium there are multiple alleles segregating in the population. In small populations, few mutations appear each generation. The ones that fix do so quickly relative to large populations. In large populations, more mutants appear over the generations. But, the ones that fix take much longer to do so. Thus, the rate of neutral evolution (in substitutions per generation) is independent of population size.

The rate of mutation determines the level of heterozygosity at a locus according to the neutral theory. Heterozygosity is simply the proportion of the population that is heterozygous. Equilibrium heterozygosity is given as H = 4Nv/[4Nv+1] (for diploid populations). H can vary from a very small number to almost one. In small populations, H is small (because the equation is approximately a very small number divided by one). In (biologically unrealistically) large populations, heterozygosity approaches one (because the equation is approximately a large number divided by itself). Directly testing this model is difficult because N and v can only be estimated for most natural populations. But, heterozygosities are believed to be too low to be described by a strictly neutral model. Solutions offered by neutralists for this discrepancy include hypothesizing that natural populations may not be at equilibrium.

At equilibrium there should be a few alleles at intermediate frequency and many at very low frequencies. This is the Ewens- Watterson distribution. New alleles enter a population every generation, most remain at low frequency until they are lost. A few drift to intermediate frequencies, a very few drift all the way to fixation. In *Drosophila pseudoobscura*, the protein Xanthine dehydrogenase (Xdh) has many variants. In a single population, Keith, et. al., found that 59 of 96 proteins were of one type, two others were represented ten and nine times and nine other types were present singly or in low numbers.

*Deleterious alleles*

Deleterious mutants are selected against but remain at low frequency in the gene pool. In diploids, a deleterious recessive mutant may increase in frequency due to drift. Selection cannot see it when it is masked by a dominant allele. Many disease causing alleles remain at low frequency for this reason. People who are carriers do not suffer the negative effect of the allele. Unless they mate with another carrier, the allele may simply continue to be passed on. Deleterious alleles also remain in populations at a low frequency due to a balance between recurrent mutation and selection. This is called the mutation load.

*Beneficial alleles*

Most new mutants are lost, even beneficial ones. Wright calculated that the probability of fixation of a beneficial allele is 2s. (This assumes a large population size, a small fitness benefit, and that heterozygotes have an intermediate fitness. A benefit of 2s yields an overall rate of evolution: k=4Nvs where v is the mutation rate to beneficial alleles) An allele that conferred a one percent increase in fitness only has a two percent chance of fixing. The probability of fixation of beneficial type of mutant is boosted by recurrent mutation. The beneficial mutant may be lost several times, but eventually it will arise and stick in a population. (Recall that even deleterious mutants recur in a population.)

Directional selection depletes genetic variation at the selected locus as the fitter allele sweeps to fixation. Sequences linked to the selected allele also increase in frequency due to hitchhiking. The lower the rate of recombination, the larger the window of sequence that hitchhikes. Begun and Aquadro compared the level of nucleotide polymorphism within and between species with the rate of recombination at a locus. Low levels of nucleotide polymorphism within species coincided with low rates of recombination. This could be explained by molecular mechanisms if recombination itself was mutagenic. In this case, recombination with also be correlated with nucleotide divergence between species. But, the level of sequence divergence did not correlate with the rate of recombination. Thus, they inferred that selection was the cause. The correlation between recombination and nucleotide polymorphism leaves the conclusion that selective sweeps occur often enough to leave an imprint on the level of genetic variation in natural populations.

One example of a beneficial mutation comes from the mosquito *Culex pipiens*. In this organism, a gene that was involved with breaking down organophosphates - common insecticide ingredients -became duplicated. Progeny of the organism with this mutation quickly swept across the worldwide mosquito population. There are numerous examples of insects developing resistance to chemicals, especially DDT which was once heavily used in this country. And, most importantly, even though "good" mutations happen much less frequently than "bad" ones, organisms with "good" mutations thrive while organisms with "bad" ones die out.

If beneficial mutants arise infrequently, the only fitness differences in a population will be due to new deleterious mutants and the deleterious recessives. Selection will simply be weeding out unfit variants. Only occasionally will a beneficial allele be sweeping through a population. The general lack of large fitness differences segregating in natural populations argues that beneficial mutants do indeed arise infrequently. However, the impact of a beneficial mutant on the level of variation at a locus can be large and lasting. It takes many generations for a locus to regain appreciable levels of heterozygosity following a selective sweep.

*Recombination*

Each chromosome in our sperm or egg cells is a mixture of genes from our mother and our father. Recombination can be thought of as gene shuffling. Most organisms have linear chromosomes and their genes lie at specific location (loci) along them. Bacteria have circular chromosomes. In most sexually reproducing organisms, there are two of each chromosome type in every cell. For instance in humans, every chromosome is paired, one inherited from the mother, the other inherited from the father. When an organism produces gametes, the gametes end up with only one of each chromosome per cell. Haploid gametes are produced from diploid cells by a process called meiosis.

In meiosis, homologous chromosomes line up. The DNA of the chromosome is broken on both chromosomes in several places and rejoined with the other strand. Later, the two homologous chromosomes are split into two separate cells that divide and become gametes. But, because of recombination, both of the chromosomes are a mix of alleles from the mother and father.

Recombination creates new combinations of alleles. Alleles that arose at different times and different places can be brought together. Recombination can occur not only between genes, but within genes as well. Recombination within a gene can form a new allele. Recombination is a mechanism of evolution because it adds new alleles and combinations of alleles to the gene pool.

### *Gene Flow*

New organisms may enter a population by migration from another population. If they mate within the population, they can bring new alleles to the local gene pool. This is called gene flow. In some closely related species, fertile hybrids can result from interspecific matings. These hybrids can vector genes from species to species.

Gene flow between more distantly related species occurs infrequently. This is called horizontal transfer. One interesting case of this involves genetic elements called P elements. Margaret Kidwell found that P elements were transferred from some species in the *Drosophila willistoni* group to *Drosophila melanogaster*. These two species of fruit flies are distantly related and hybrids do not form. Their ranges do, however, overlap. The P elements were vectored into *D. melanogaster* via a parasitic mite that targets both these species. This mite punctures the exoskeleton of the flies and feeds on the "juices". Material, including DNA, from one fly can be transferred to another when the mite feeds. Since P elements actively move in the genome (they are themselves parasites of DNA), one incorporated itself into the genome of a *melanogaster* fly and subsequently spread through the species. Laboratory stocks of *melanogaster* caught prior to the 1940's lack of P elements. All natural populations today harbor them.

### *Overview of Evolution within a Lineage*

Evolution is a change in the gene pool of a population over time; it can occur due to several factors. Three mechanisms add new alleles to the gene pool: mutation, recombination and gene flow. Two mechanisms remove alleles, genetic drift and natural selection. Drift removes alleles randomly from the gene pool. Selection removes deleterious alleles from the gene pool. The amount of genetic variation found in a population is the balance between the actions of these mechanisms.

Natural selection can also increase the frequency of an allele. Selection that weeds out harmful alleles is called negative selection. Selection that increases the frequency of helpful alleles is called positive, or sometimes positive Darwinian, selection. A new allele can also drift to high frequency. But, since the change in frequency of an allele each generation is random, nobody speaks of positive or negative drift.

Except in rare cases of high gene flow, new alleles enter the gene pool as a single copy. Most new alleles added to the gene pool are lost almost immediately due to drift or selection; only a small percent ever reach a high frequency in the population. Even most moderately beneficial alleles are lost due to drift when they appear. But, a mutation can reappear numerous times.

The fate of any new allele depends a great deal on the organism it appears in. This allele will be linked to the other alleles near it for many generations. A mutant allele can increase in frequency simply because it is linked to a beneficial allele at a nearby locus. This can occur even if the mutant allele is deleterious, although it must not be so deleterious as to offset the benefit of the other allele. Likewise a potentially beneficial new allele can be eliminated from the gene pool because it was linked to deleterious alleles when it first arose. An allele "riding on the coat tails" of a beneficial allele is called a hitchhiker. Eventually, recombination will bring the two loci to linkage equilibrium. But, the more closely linked two alleles are, the longer the hitchhiking will last.

The effects of selection and drift are coupled. Drift is intensified as selection pressures increase. This is because increased selection (i.e. a greater difference in reproductive success among organisms in a population) reduces the effective population size, the number of individuals contributing alleles to the next generation.

Adaptation is brought about by cumulative natural selection, the repeated sifting of mutations by natural selection. Small changes, favored by selection, can be the stepping-stone to further changes. The summation of large numbers of these changes is macroevolution.

### *The Development of Evolutionary Theory*

Biology came of age as a science when Charles Darwin published "On the Origin of Species." But, the idea of evolution wasn't new to Darwin. Lamarck published a theory of evolution in 1809. Lamarck thought that species arose continually from nonliving sources. These species were initially very primitive, but increased in complexity over time due to some inherent tendency. This type of evolution is called orthogenesis. Lamarck proposed that an organism's acclimation to the environment could be passed on to its offspring. For example, he thought proto-giraffes stretched their necks to reach higher twigs. This caused their offspring to be born with longer necks. This proposed mechanism of evolution is called the inheritance of acquired characteristics. Lamarck also believed species never went extinct, although they may change into newer forms. All three of these ideas are now known to be wrong.

Darwin's contributions include hypothesizing the pattern of common descent and proposing a mechanism for evolution — natural selection. In Darwin's theory of natural selection, new variants arise continually within populations. A small percentage of these variants cause their bearers to produce more offspring than others. These variants thrive and supplant their less productive competitors. The effect of numerous instances of selection would lead to a species being modified over time.

Darwin's theory did not accord with older theories of genetics. In Darwin's time, biologists held to the theory of blending inheritance — an offspring was an average of its parents. If an individual had one short parent and one tall parent, it would be of medium height. And, the offspring would pass on genes for medium sized offspring. If this was the case, new genetic variations would quickly be diluted out of a population. They could not accumulate as the theory of evolution required. We now know that the idea of blending inheritance is wrong.

Darwin didn't know that the true mode of inheritance was discovered in his lifetime. Gregor Mendel, in his experiments on hybrid peas, showed that genes from a mother and father do not blend. An offspring from a short and a tall parent may be medium sized; but it carries genes for shortness and tallness. The genes remain distinct and can be passed on to subsequent generations. Mendel mailed his paper to Darwin, but Darwin never opened it.

It was a long time until Mendel's ideas were accepted. One group of biologists, called biometricians, thought Mendel's laws only held for a few traits. Most traits, they claimed, were governed by blending inheritance. Mendel studied discrete traits. Two of the traits in his famous experiments were smooth versus wrinkled coat on peas. This trait did not vary continuously. In other words, peas are either wrinkled or smooth — intermediates are not found. Biometricians considered these traits aberrations. They studied continuously varying traits like size and believed most traits showed blending inheritance.

### *Incorporating Genetics into Evolutionary Theory*

The discrete genes Mendel discovered would exist at some frequency in natural populations. Biologists wondered how and if these frequencies would change. Many thought that the more common versions of genes would increase in frequency simply because they were already at high frequency.

Hardy and Weinberg independently showed that the frequency of an allele would not change over time simply due to its being rare or common. Their model had several assumptions — that all alleles reproduced at the same rate, that the population size was very large and that alleles did not change in form. Later, R. A. Fisher showed that Mendel's laws could explain continuous traits if the expression of these traits were due to the action of many genes. After this, geneticists accepted Mendel's Laws as the basic rules of genetics. From this basis, Fisher, Sewall Wright and J. B. S.. Haldane founded the field of population genetics. Population genetics is a field of biology that attempts to measure and explain the levels of genetic variation in populations.

R. A. Fisher studied the effect of natural selection on large populations. He demonstrated that even very small selective differences amongst alleles could cause appreciable changes in allele frequencies over time. He also showed that the rate of adaptive change in a population is proportional to the amount of genetic variation present. This is called Fisher's Fundamental Theorem of Natural Selection. Although it is called the fundamental theorem, it does not hold in all cases. The rate at which natural selection brings about adaptation depends on the details of how selection is working. In some rare cases, natural selection can actually cause a decline in the mean relative fitness of a population.

Sewall Wright was more concerned with drift. He stressed that large populations are often subdivided into many subpopulations. In his theory, genetic drift played a more important role compared to selection. Differentiation between subpopulations, followed by migration among them, could contribute to adaptations amongst populations. Wright also came up with the idea of the adaptive landscape — an idea that remains influential to this day. Its influence remains even though P. A. P. Moran has shown that, mathematically, adaptive landscapes don't exist as Wright envisioned them. Wright extended his results of one-locus models to a two-locus case in proposing the adaptive landscape. But, unbeknownst to him, the general conclusions of the one-locus model don't extend to the two-locus case.

J. B. S. Haldane developed many of the mathematical models of natural and artificial selection. He showed that selection and mutation could oppose each other, that deleterious mutations could remain in a population due to recurrent mutation. He also demonstrated that there was a cost to natural selection, placing a limit on the amount of adaptive substitutions a population could undergo in a given time frame.

For a long time, population genetics developed as a theoretical field. But, gathering the data needed to test the theories was nearly impossible. Prior to the advent of molecular biology, estimates of genetic variability could only be inferred from levels of morphological differences in populations. Lewontin and Hubby were the first to get a good estimate of genetic variation in a population. Using the then new technique of protein electrophoresis, they showed that 30% of the loci in a population of *Drosophila pseudoobscura* were polymorphic. They also showed that it was likely that they could not detect all the variation that was present. Upon finding this level of variation, the question became — was this maintained by natural selection, or simply the result of genetic drift? This level of variation was too high to be explained by balancing selection.

Motoo Kimura theorized that most variation found in populations was selectively equivalent (neutral). Multiple alleles at a locus differed in sequence, but their fitnesses were the same. Kimura's neutral theory described rates of evolution and levels of polymorphism solely in terms of mutation and genetic drift. The neutral theory did not deny that natural selection acted on natural populations; but it claimed that the majority of natural variation was transient polymorphisms of neutral alleles. Selection did not act frequently or strongly enough to influence rates of evolution or levels of polymorphism.

Initially, a wide variety of observations seemed to be consistent with the neutral theory. Eventually, however, several lines of evidence toppled it. There is less variation in natural populations than the neutral theory predicts. Also, there is too much variance in rates of substitutions in different lineages to be explained by mutation and drift alone. Finally, selection itself has been shown to have an impact on levels of nucleotide variation. Currently, there is no comprehensive mathematical theory of evolution that accurately predicts rates of evolution and levels of heterozygosity in natural populations.

### *Evolution Among Lineages*

#### *The Pattern of Macroevolution*

Evolution is not progress. The popular notion that evolution can be represented as a series of improvements from simple cells, through more complex life forms, to humans (the pinnacle of evolution), can be traced to the concept of the scale of nature. This view is incorrect. All species have descended from a common ancestor. As time went on, different lineages of organisms were modified with descent to adapt to their environments. Thus, evolution is best viewed as a branching tree or bush, with the tips of each branch representing currently living species. No living organisms today are our ancestors. Every living species is as fully modern as we are with its own unique evolutionary history. No extant species are "lower life forms," atavistic stepping stones paving the road to humanity.

A related, and common, fallacy about evolution is that humans evolved from some living species of ape. This is not the case — humans and apes share a common ancestor. Both humans and living apes are fully modern species; the ancestor we evolved from was an ape, but it is now extinct and was not the same as present day apes (or humans for that matter). If it were not for the vanity of human beings, we would be classified as an ape. Our closest relatives are, collectively, the chimpanzee and the pygmy chimp. Our next nearest relative is the gorilla.

*Evidence for Common Descent and Macroevolution*

Microevolution can be studied directly. Macroevolution cannot. Macroevolution is studied by examining patterns in biological populations and groups of related organisms and inferring process from pattern. Given the observation of microevolution and the knowledge that the earth is billions of years old — macroevolution could be postulated. But this extrapolation, in and of itself, does not provide a compelling explanation of the patterns of biological diversity we see today. Evidence for macroevolution, or common ancestry and modification with descent, comes from several other fields of study. These include: comparative biochemical and genetic studies, comparative developmental biology, patterns of biogeography, comparative morphology and anatomy and the fossil record.

Closely related species (as determined by morphologists) have similar gene sequences. Overall sequence similarity is not the whole story, however. The pattern of differences we see in closely related genomes is worth examining. All living organisms use DNA as their genetic material, although some viruses use RNA. DNA is composed of strings of nucleotides. There are four different kinds of nucleotides: adenine (A), guanine (G), cytosine (C) and thymine (T). Genes are sequences of nucleotides that code for proteins. Within a gene, each block of three nucleotides is called a codon. Each codon designates an amino acid (the subunits of proteins).

The three letter code is the same for all organisms (with a few exceptions). There are 64 codons, but only 20 amino acids to code for; so, most amino acids are coded for by several codons. In many cases the first two nucleotides in the codon designate the amino acid. The third position can have any of the four nucleotides and not effect how the code is translated.

A gene, when in use, is transcribed into RNA — a nucleic acid similar to DNA. (RNA, like DNA, is made up of nucleotides although t he nucleotide uracil (U) is used in place of thymine (T).) The RNA transcribed from a gene is called messenger RNA. Messenger RNA is then translated via cellular machinery called ribosomes into a string of amino acids — a protein. Some proteins function as enzymes, catalysts that speed the chemical reactions in cells. Others are structural or involved in regulating development.

Gene sequences in closely related species are very similar. Often, the same codon specifies a given amino acid in two related species, even though alternate codons could serve functionally as well. But, some differences do exist in gene sequences. Most often, differences are in third codon positions, where changes in the DNA sequence would not disrupt the sequence of the protein.

There are other sites in the genome where nucleotide differences do not effect protein sequences. The genome of eukaryotes is loaded with 'dead genes' called pseudogenes. Pseudogenes are copies of working genes that have been inactivated by mutation. Most pseudogenes do not produce full proteins. They may be transcribed, but not translated. Or, they may be translated, but only a truncated protein is produced. Pseudogenes evolve much faster than their working counterparts. Mutations in them do not get incorporated into proteins, so they have no effect on the fitness of an organism. Introns are sequences of DNA that interrupt a gene, but do not code for anything. The coding portions of a gene are called exons. Introns are spliced out of the messenger RNA prior to translation, so they do not contribute information needed to make the protein. They are sometimes, however, involved in regulation of the gene. Like pseudogenes, introns (in general) evolve faster than coding portions of a gene.

Nucleotide positions that can be changed without changing the sequence of a protein are called silent sites. Sites where changes result in an amino acid substitution are called replacement sites. Silent sites are expected to be more polymorphic within a population and show more differences between populations. Although both silent and replacement sites receive the same amount of mutations, natural selection only infrequently allows changes at replacement sites. Silent sites, however, are not as constrained.

Kreitman was the first demonstrate that silent sites

were more variable than coding sites. Shortly after the methods of DNA sequencing were discovered, he sequenced 11 alleles of the enzyme alcohol dehydrogenase (AdH). Of the 43 polymorphic nucleotide sites he found, only one resulted in a change in the amino acid sequence of the protein.

Silent sites may not be entirely selectively neutral. Some DNA sequences are involved with regulation of genes, changes in these sites may be deleterious. Likewise, although several codons code for a single amino acid, an organism may have a preferred codon for each amino acid. This is called codon bias.

If two species shared a recent common ancestor one would expect genetic information, even information such as redundant nucleotides and the position of introns or pseudogenes, to be similar. Both species would have inherited this information from their common ancestor. The degree of similarity in nucleotide sequence is a function of divergence time. If two populations had recently separated, few differences would have built up between them. If they separated long ago, each population would have evolved numerous differences from their common ancestor (and each other). The degree of similarity would also be a function of silent versus replacement sites. Li and Graur, in their molecular evolution text, give the rates of evolution for silent vs. replacement rates. The rates were estimated from sequence comparisons of 30 genes from humans and rodents, which diverged about 80 million years ago. Silent sites evolved at an average rate of 4.61 nucleotide substitution per $10^9$ years. Replacement sites evolved much slower at an average rate of 0.85 nucleotide substitutions per $10^9$ years.

Groups of related organisms are 'variations on a theme' — the same set of bones are used to construct all vertebrates.

The bones of the human hand grow out of the same tissue as the bones of a bat's wing or a whale's flipper; and, they share many identifying features such as muscle insertion points and ridges. The only difference is that they are scaled differently. Evolutionary biologists say this indicates that all mammals are modified descendants of a common ancestor which had the same set of bones.

Closely related organisms share similar developmental pathways. The differences in development are most evident at the end. As organisms evolve, their developmental pathway gets modified. An alteration near the end of a developmental pathway is less likely to be deleterious than changes in early development. Changes early on may have a cascading effect. Thus most evolutionary changes in development are expected to take place at the periphery of development, or in early aspects of development that have no later repercussions. For a change in early development to be propagated, the benefit of the early alteration must outweigh the consequences to later development.

Because they have evolved this way, organisms pass through the early stages of development that their ancestors passed through up to the point of divergence. So, an organism's development mimics its ancestors although it doesn't recreate it exactly. Development of the flatfish, Pleuronectes, illustrates this point. Early on, Pleuronectes develops a tail that comes to a point. In the next developmental stage, the top lobe of the tail is larger than the bottom lobe (as in sharks). When development is complete, the upper and lower lobes are equally sized. This developmental pattern mirrors the evolutionary transitions it has undergone.

Natural selection can modify any stage of a life cycle, so some differences are seen in early development. Thus, evolution does not always recapitulate ancestral forms — butterflies did not evolve from ancestral caterpillars, for example. There are differences in the appearance of early vertebrate embryos. Amphibians rapidly form a ball of cells in early development. Birds, reptiles and mammals form a disk. The shape of the early embryo is a result of different yolk concentrations in the eggs. Birds' and reptiles' eggs are heavily yolked. Their eggs develop similarly to amphibians except the yolk has deformed the shape of the embryo. The ball is stretched out and lying atop the yolk. Mammals have no yolk, but still form a disk early. This is because they have descended from reptiles. Mammals lost their yolky eggs, but retained the early pattern of development. In all these vertebrates, the pattern of cell movements is similar despite superficial differences in appearance. In addition, all types quickly converge upon a primitive, fish-like stage within a few days. From there, development diverges.

Traces of an organism's ancestry sometimes remain even when an organism's development is complete. These are called vestigial structures. Many snakes have rudimentary pelvic bones retained from their walking ancestors. Vestigial does not mean useless, it means the structure is clearly a vestige of an structure inherited from ancestral organisms. Vestigial structures may acquire new functions. In humans, the appendix now houses some immune system cells.

Closely related organisms are usually found in close geographic proximity; this is especially true of organisms with limited dispersal opportunities. The mammalian fauna of Australia is often cited as an example of this; marsupial mammals fill most of the equivalent niches that placentals fill in other ecosystems. If all organisms descended from a common ancestor, species distribution across the planet would be a function of site of origination, potential for dispersal, distribution of suitable habitat, and time since origination. In the case of Australian mammals, their physical separation from sources of placentals means potential niches were filled by a marsupial radiation rather than a placental radiation or invasion.

Natural selection can only mold available genetically based variation. In addition, natural selection provides no mechanism for advance planning. If selection can only tinker with the available genetic variation, we should expect to see examples of jury-rigged design in living species. This is indeed the case. In lizards of the genus *Cnemodophorus*, females reproduce parthenogenetically. Fertility in these lizards is increased when a female mounts another female and simulates copulation. These lizards evolved from sexual lizards whose hormones were aroused by sexual behavior. Now, although the sexual mode of reproduction has been lost, the means of getting aroused (and hence fertile) has been retained.

Fossils show hard structures of organisms less and less similar to modern organisms in progressively older rocks. In addition, patterns of biogeography apply to fossils as well as extant organisms. When combined with plate tectonics, fossils provide evidence of distributions and dispersals of ancient species. For example, South America had a very distinct marsupial mammalian fauna until the land bridge formed between North and South America. After that marsupials started disappearing and placentals took their place. This is commonly interpreted as the placentals wiping out the marsupials, but this may be an over simplification.

Transitional fossils between groups have been found. One of the most impressive transitional series is the ancient reptile to modern mammal transition. Mammals and reptiles differ in skeletal details, especially in their skulls. Reptilian jaws have four bones. The foremost is called the dentary. In mammals, the dentary bone is the only bone in the lower jaw. The other bones are part of the middle ear. Reptiles have a weak jaw and a mouthful of undifferentiated teeth. Their jaw is closed by three muscles: the external, posterior and internal adductor. Each reptile tooth is single cusped. Mammals have powerful jaws with differentiated teeth. Many of these teeth, such as the molars, are multi-cusped. The temporalis and masseter muscles, derived from the external adductor, close the mammalian jaw. Mammals have a secondary palate, a bony structure separating their nostril passages and throat, so most can swallow and breathe simultaneously. Reptiles lack this.

The evolution of these traits can be seen in a series of fossils. *Procynosuchus* shows an increase in size of the dentary bone and the beginnings of a palate. *Thrinaxodon* has a reduced number of incisors, a precursor to tooth differentiation. *Cynognathus* (a doglike carnivore) shows a further increase in size of the dentary bone. The other three bones are located inside the back portion of the jaw. Some teeth are multicusped and the teeth fit together tightly. *Diademodon* (a plant eater) shows a more advanced degree of occlusion (teeth fitting tightly). *Probelesodon* has developed a double joint in the jaw. The jaw could hinge off two points with the upper skull. The front hinge was probably the actual hinge while the rear hinge was an alignment guide. The forward movement of a hinge point allowed for the precursor to the modern masseter muscle to anchor further forward in the jaw. This allowed for a more powerful bite. The first true mammal was *Morgonucudon*, a rodent-like insectivore from the late Triassic. It had all the traits common to modern mammals. These species were not from a single, unbranched lineage. Each is an example from a group of organisms along the main line of mammalian ancestry.

The strongest evidence for macroevolution comes from the fact that suites of traits in biological entities fall into a nested pattern. For example, plants can be divided into two broad categories, non- vascular (ex. mosses) and vascular. Vascular plants can be divided into seedless (ex. ferns) and seeded. Vascular seeded plants can be divided into gymnosperms (ex. pines) and flowering plants (angiosperms). Angiosperms can be divided into monocots and dicots. Each of these types of plants have several characters that distinguish them from other plants. Traits are not mixed and matched in groups of organisms. For example, flowers are only seen in plants that carry several other characters that distinguish them as angiosperms. This is the expected pattern of common descent. All the species in a group will share traits they inherited from their common ancestor. But, each subgroup will have evolved unique traits of its own. Similarities bind groups together. Differences show how they are subdivided.

The real test of any scientific theory is its ability to generate testable predictions and, of course, have the predictions borne out. Evolution easily meets this criterion. In several of the above examples I stated, closely related organisms share X. If I define closely related as sharing X, this is an empty statement. It does however, provide a prediction. If two organisms share a similar anatomy, one would then predict that their gene sequences would be more similar than a morphologically distinct organism. This has been spectacularly borne out by the recent flood of gene sequences — the correspondence to trees drawn by morphological data is very high. The discrepancies are never too great and usually confined to cases where the pattern of relationship was debated.

*Mechanisms of Macroevolution*

The following deals with mechanisms of evolution above the species level.

*Speciation — Increasing Biological Diversity*

Speciation is the process of a single species becoming two or more species. Many biologists think speciation is key to understanding evolution. Some would argue that certain evolutionary phenomena apply only at speciation and macroevolutionary change cannot occur without speciation. Other biologists think major evolutionary change can occur without speciation. Changes between lineages are only an extension of the changes within each lineage. In general, paleontologists fall into the former category and geneticists in the latter.

***Modes of Speciation***

Biologists recognize two types of speciation: allopatric and sympatric speciation. The two differ in geographical distribution of the populations in question. Allopatric speciation is thought to be the most common form of speciation. It occurs when a population is split into two (or more) geographically isolated subdivisions that organisms cannot bridge. Eventually, the two populations' gene pools change independently until they could not interbreed even if they were brought back together. In other words, they have speciated.

Sympatric speciation occurs when two subpopulations become reproductively isolated without first becoming geographically isolated. Insects that live on a single host plant provide a model for sympatric speciation. If a group of insects switched host plants they would not breed with other members of their species still living on their former host plant. The two subpopulations could diverge and speciate. Agricultural records show that a strain of the apple maggot fly *Rhagolettis pomenella* began infesting apples in the 1860's. Formerly it had only infested hawthorn fruit. Feder, Chilcote and Bush have shown that two races of *Rhagolettis pomenella* have become behaviorally isolated. Allele frequencies at six loci (aconitase 2, malic enzyme, mannose phosphate isomerase, aspartate amino-transferase, NADH-diaphorase-2, and beta-hydroxy acid dehydrogenase) are diverging. Significant amounts of linkage disequilibrium have been found at these loci, indicating that they may all be hitchhiking on some allele under selection. Some biologists call sympatric speciation microallopatric speciation to emphasize that the subpopulations are still physically separate at an ecological level.

Biologists know little about the genetic mechanisms of speciation. Some think a series of small changes in each subdivision gradually lead to speciation. The founder effect could set the stage for relatively rapid speciation, a genetic revolution in Ernst Mayr's terms. Alan Templeton hypothesized that a few key genes could change and confer reproductive isolation. He called this a genetic transilience. Lynn Margulis thinks most speciation events are caused by changes in internal symbionts. Populations of organisms are very complicated. It is likely that there are many ways speciation can occur. Thus, all of the above ideas may be correct, each in different circumstances. Darwin's book was titled "The Origin of Species" despite the fact that he did not really address this question; over one hundred and fifty years later, how species originate is still largely a mystery.

*Observed Speciations*

Speciation has been observed. In the plant genus *Tragopogon*, two new species have evolved within the past 50-60 years. They are *T. mirus* and *T. miscellus*. The new species were formed when one diploid species fertilized a different diploid species and produced a tetraploid offspring.

This tetraploid offspring could not fertilize or be fertilized by either of its two parent species types. It is reproductively isolated, the definition of a species.

### *Extinction — Decreasing Biological Diversity*

#### *Ordinary Extinction*

Extinction is the ultimate fate of all species. The reasons for extinction are numerous. A species can be competitively excluded by a closely related species, the habitat a species lives in can disappear and/or the organisms that the species exploits could come up with an unbeatable defense.

Some species enjoy a long tenure on the planet while others are short- lived. Some biologists believe species are programmed to go extinct in a manner analogous to organisms being destined to die. The majority, however, believe that if the environment stays fairly constant, a well adapted species could continue to survive indefinitely.

#### *Mass Extinction*

Mass extinctions shape the overall pattern of macroevolution. If you view evolution as a branching tree, it's best to picture it as one that has been severely pruned a few times in its life. The history of life on this earth includes many episodes of mass extinction in which many groups of organisms were wiped off the face of the planet. Mass extinctions are followed by periods of radiation where new species evolve to fill the empty niches left behind. It is probable that surviving a mass extinction is largely a function of luck. Thus, contingency plays a large role in patterns of macroevolution.

The largest mass extinction came at the end of the Permian, about 250 million years ago. This coincides with the formation of Pangaea II, when all the world's continents were brought together by plate tectonics. A worldwide drop in sea level also occurred at this time.

The most well-known extinction occurred at the boundary between the Cretaceous and Tertiary Periods. This called the K/T Boundary and is dated at around 65 million years ago. This extinction eradicated the dinosaurs. The K/T event was probably caused by environmental disruption brought on by a large impact of an asteroid with the earth. Following this extinction the mammalian radiation occurred. Mammals coexisted for a long time with the dinosaurs but were confined mostly to nocturnal insectivore niches. With the eradication of the dinosaurs, mammals radiated to fill the vacant niches.

Currently, human alteration of the ecosphere is causing a global mass extinction.

*Punctuated Equilibrium*

The theory of punctuated equilibrium is an inference about the process of macroevolution from the pattern of species documented in the fossil record. In the fossil record, transition from one species to another is usually abrupt in most geographic locales — no transitional forms are found. In short, it appears that species remain unchanged for long stretches of time and then are quickly replaced by new species. However, if wide ranges are searched, transitional forms that bridge the gap between the two species are sometimes found in small, localized areas. For example, in Jurassic brachiopods of the genus *Kutchithyris*, *K. acutiplicata* appears below another species, *K. euryptycha*. Both species were common and covered a wide geographical area. They differ enough that some have argued they should be in a different genera. In just one small locality an approximately 1.25m sedimentary layer with these fossils is found. In the narrow (10 cm) layer that separates the two species, both species are found along with transitional forms. In other localities there is a sharp transition.

Eldredge and Gould proposed that most major morphological change occurs (relatively) quickly in small peripheral population at the time of speciation. New forms will then invade the range of their ancestral species. Thus, at most locations that fossils are found, transition from one species to another will be abrupt. This abrupt change will reflect replacement by migration however, not evolution. In order to find the transitional fossils, the area of speciation must be found.

There has been considerable confusion about the theory. Some popular accounts give the impression that abrupt changes in the fossil record are due to blindingly fast evolution; this is not a part of the theory.

Punctuated equilibrium has been presented as a hierarchical theory of evolution. Proponents of punctuated equilibrium see speciation as analogous to mutation and the replacement of one species by another as analogous to natural selection. This is called species selection. Speciation adds new species to the species pool just as mutation adds new alleles to the gene pool. Species selection favors one species over another just as natural selection can favor one allele over another. Evolutionary trends within a group would be the result of selection among species, not natural selection acting within species. This is the most controversial part of the theory. Many biologists agree with the pattern of macroevolution these paleontologists posit, but believe species selection is not even theoretically likely to occur.

Critics would argue that species selection is not analogous to natural selection and therefore evolution is not hierarchical. Also, the number of species produced over time is far less than the amount of different alleles that enter gene pools over time. So, the amount of adaptive evolution produced by species selection (if it did occur) would have to be orders of magnitude less than adaptive evolution within populations by natural selection.

Tests of punctuated equilibrium have been equivocal. It has been known for a long time that rates of evolution vary over time, that is not controversial. However, phylogenetic studies conflict as to whether there is a clear association between speciation and morphological change. In addition, there are major polymorphisms within some species. For example, bluegill sunfish have two male morphs. One is a large, long-lived, mate-protecting male; the other is a smaller, shorter-lived male who sneaks matings from females guarded by large males. The existence of within species polymorphisms demonstrates that speciation is not a requirement for major morphological change.

### *A Brief History of Life*

Biologists studying evolution do a variety of things: population geneticists study the process as it is occurring; systematists seek to determine relationships between species and paleontologists seek to uncover details of the unfolding of life in the past. Discerning these details is often difficult, but hypotheses can be made and tested as new evidence comes to light. This section should be viewed as the best hypothesis scientists have as to the history of the planet. The material here ranges from some issues that are fairly certain to some topics that are nothing more than informed speculation. For some points there are opposing hypotheses — I have tried to compile a consensus picture. In general, the more remote the time, the more likely the story is incomplete or in error.

Life evolved in the sea. It stayed there for the majority of the history of earth. The first replicating molecules were most likely RNA. RNA is a nucleic acid similar to DNA. In laboratory studies it has been shown that some RNA sequences have catalytic capabilities. Most importantly, certain RNA sequences act as polymerases — enzymes that form strands of RNA from its monomers. This process of self replication is the crucial step in the formation of life. This is called the RNA world hypothesis.

The common ancestor of all life probably used RNA as its genetic material. This ancestor gave rise to three major lineages of life. These are: the prokaryotes ("ordinary" bacteria), archaebacteria (thermophilic, methanogenic and halophilic bacteria) and eukaryotes. Eukaryotes include protists (single celled organisms like amoebas and diatoms and a few multicellular forms such as kelp), fungi (including mushrooms and yeast), plants and animals. Eukaryotes and archaebacteria are the two most closely related of the three. The process of translation (making protein from the instructions on a messenger RNA template) is similar in these lineages, but the organization of the genome and transcription (making messenger RNA from a DNA template) is very different in prokaryotes than in eukaryotes and archaebacteria. Scientists interpret this to mean that the common ancestor was RNA based; it gave rise to two lineages that independently formed a DNA genome and hence independently evolved mechanisms to transcribe DNA into RNA.

The first cells must have been anaerobic because there was no oxygen in the atmosphere. In addition, they were probably thermophilic ("heat-loving") and fermentative. Rocks as old as 3.5 billion years old have yielded prokaryotic fossils. Specifically, some rocks from Australia called the Warrawoona series give evidence of bacterial communities organized into structures called stromatolites. Fossils like these have subsequently been found all over the world. These mats of bacteria still form today in a few locales (for example, Shark Bay Australia). Bacteria are the only life forms found in the rocks for a long, long time —eukaryotes (protists) appear about 1.5 billion years ago and fungi-like things appear about 900 million years ago (0.9 billion years ago).

Photosynthesis evolved around 3.4 billion years ago. Photosynthesis is a process that allows organisms to harness sunlight to manufacture sugar from simpler precursors. The first photosystem to evolve, PSI, uses light to convert carbon dioxide (CO2) and hydrogen sulfide (H2S) to glucose. This process releases sulfur as a waste product. About a billion years later, a second photosystem (PS) evolved, probably from a duplication of the first photosystem. Organisms with PSII use both photosystems in conjunction to convert carbon dioxide (CO2) and water (H2O) into glucose. This process releases oxygen as a waste product. Anoxygenic (or H2S) photosynthesis, using PSI, is seen in living purple and green bacteria. Oxygenic (or H2O) photosynthesis, using PSI and PSII, takes place in cyanobacteria. Cyanobacteria are closely related to and hence probably evolved from purple bacterial ancestors. Green bacteria are an outgroup. Since oxygenic bacteria are a lineage within a cluster of anoxygenic lineages, scientists infer that PSI evolved first. This also corroborates with geological evidence.

Green plants and algae also use both photosystems. In these organisms, photosynthesis occurs in organelles (membrane bound structures within the cell) called chloroplasts. These organelles originated as free living bacteria related to the cyanobacteria that were engulfed by ur-eukaryotes and eventually entered into an endosymbiotic relationship. This endosymbiotic theory of eukaryotic organelles was championed by Lynn Margulis. Originally controversial, this theory is now accepted. One key line of evidence in support of this idea came when the DNA inside chloroplasts was sequenced — the gene sequences were more similar to free-living cyanobacteria sequences than to sequences from the plants the chloroplasts resided in.

After the advent of photosystem II, oxygen levels increased. Dissolved oxygen in the oceans increased as well as atmospheric oxygen. This is sometimes called the oxygen holocaust. Oxygen is a very good electron acceptor and can be very damaging to living organisms. Many bacteria are anaerobic and die almost immediately in the presence of oxygen. Other organisms, like animals, have special ways to avoid cellular damage due to this element (and in fact require it to live.) Initially, when oxygen began building up in the environment, it was neutralized by materials already present. Iron, which existed in high concentrations in the sea was oxidized and precipitated. Evidence of this can be seen in banded iron formations from this time, layers of iron deposited on the sea floor. As one geologist put it, "the world rusted." Eventually, it grew to high enough concentrations to be dangerous to living things. In response, many species went extinct, some continued (and still continue) to thrive in anaerobic microenvironments and several lineages independently evolved oxygen respiration.

The purple bacteria evolved oxygen respiration by reversing the flow of molecules through their carbon fixing pathways and modifying their electron transport chains. Purple bacteria also enabled the eukaryotic lineage to become aerobic. Eukaryotic cells have membrane bound organelles called mitochondria that take care of respiration for the cell. These are endosymbionts like chloroplasts. Mitochondria formed this symbiotic relationship very early in eukaryotic history, all but a few groups of eukaryotic cells have mitochondria. Later, a few lineages picked up chloroplasts. Chloroplasts have multiple origins. Red algae picked up ur-chloroplasts from the cyanobacterial lineage. Green algae, the group plants evolved from, picked up different urchloroplasts from a prochlorophyte, a lineage closely related to cyanobacteria.

Animals start appearing prior to the Cambrian, about 600 million years ago. The first animals dating from just before the Cambrian were found in rocks near Adelaide, Australia. They are called the Ediacarian fauna and have subsequently been found in other locales as well. It is unclear if these forms have any surviving descendants. Some look a bit like Cnidarians (jellyfish, sea anemones and the like); others resemble annelids (earthworms). All the phyla (the second highest taxonomic category) of animals appeared around the Cambrian. The Cambrian 'explosion' may have been a result of higher oxygen concentrations enabling larger organisms with higher metabolisms to evolve. Or it might be due to the spreading of shallow seas at that time providing a variety of new niches. In any case, the radiation produced a wide variety of animals.

Some paleontologists think more animal phyla were present then than now. The animals of the Burgess shale are an example of Cambrian animal fossils. These fossils, from Canada, show a bizarre array of creatures, some which appear to have unique body plans unlike those seen in any living animals.

The extent of the Cambrian explosion is often overstated. Although quick, the Cambrian explosion is not instantaneous in geologic time. Also, there is evidence of animal life prior to the Cambrian. In addition, although all the phyla of animals came into being, these were *not* the modern forms we see today. Our own phylum (which we share with other mammals, reptiles, birds, amphibians and fish) was represented by a small, sliver-like thing called *Pikaia*. Plants were not yet present. Photosynthetic protists and algae were the bottom of the food chain. Following the Cambrian, the number of marine families leveled off at a little less than 200.

The Ordovician explosion, around 500 million years ago, followed. This 'explosion', larger than the Cambrian, introduced numerous families of the Paleozoic fauna (including crinoids, articulate brachiopods, cephalopods and corals). The Cambrian fauna, (trilobites, inarticulate brachiopods, etc.) declined slowly during this time. By the end of the Ordovician, the Cambrian fauna had mostly given way to the Paleozoic fauna and the number of marine families was just over 400. It stayed at this level until the end of the Permian period.

Plants evolved from ancient green algae over 400 million years ago. Both groups use chlorophyll a and b as photosynthetic pigments. In addition, plants and green algae are the only groups to store starch in their chloroplasts. Plants and fungi (in symbiosis) invaded the land about 400 million years ago. The first plants were moss-like and required moist environments to survive. Later, evolutionary developments such as a waxy cuticle allowed some plants to exploit more inland environments. Still mosses lack true vascular tissue to transport fluids and nutrients. This limits their size since these must diffuse through the plant. Vascular plants evolved from mosses. The first vascular land plant known is *Cooksonia*, a spiky, branching, leafless structure. At the same time, or shortly thereafter, arthropods followed plants onto the land. The first land animals known are myriapods — centipedes and millipedes.

Vertebrates moved onto the land by the Devonian period, about 380 million years ago. *Ichthyostega*, an amphibian, is the among the first known land vertebrates. It was found in Greenland and was derived from lobe-finned fishes called Rhipidistians. Amphibians gave rise to reptiles. Reptiles had evolved scales to decrease water loss and a shelled egg permitting young to be hatched on land. Among the earliest well preserved reptiles is *Hylonomus*, from rocks in Nova Scotia.

The Permian extinction was the largest extinction in history. It happened about 250 million years ago. The last of the Cambrian Fauna went extinct. The Paleozoic fauna took a nose dive from about 300 families to about 50. It is estimated that 96% of all species (50% of all Families) met their end. Following this event, the Modern fauna, which had been slowly expanding since the Ordovician, took over.

The Modern fauna includes fish, bivalves, gastropods and crabs. These were barely affected by the Permian extinction. The Modern fauna subsequently increased to over 600 marine families at present. The Paleozoic fauna held steady at about 100 families. A second extinction event shortly following the Permian kept animal diversity low for awhile.

During the Carboniferous (the period just prior to the Permian) and in the Permian the landscape was dominated by ferns and their relatives. After the Permian extinction, gymnosperms (ex. pines) became more abundant. Gymnosperms had evolved seeds, from seedless fern ancestors, which helped their ability to disperse. Gymnosperms also evolved pollen, encased sperm which allowed for more outcrossing.

Dinosaurs evolved from archosaur reptiles, their closest living relatives are crocodiles. One modification that may have been a key to their success was the evolution of an upright stance. Amphibians and reptiles have a splayed stance and walk with an undulating pattern because their limbs are modified from fins. Their gait is modified from the swimming movement of fish. Splay stanced animals cannot sustain continued locomotion because they cannot breathe while they move; their undulating movement compresses their chest cavity. Thus, they must stop every few steps and breath before continuing on their way. Dinosaurs evolved an upright stance similar to the upright stance mammals independently evolved. This allowed for continual locomotion. In addition, dinosaurs evolved to be warm-blooded. Warmbloodedness allows an increase in the vigor of movements in erect organisms. Splay stanced organisms would probably not benefit from warm-bloodedness. Birds evolved from sauriscian dinosaurs. Cladistically, birds are dinosaurs. The transitional fossil *Archaeopteryx* has a mixture of reptilian and avian features.

Angiosperms evolved from gymnosperms, their closest relatives are Gnetae. Two key adaptations allowed them to displace gymnosperms as the dominant fauna — fruits and flowers. Fruits (modified plant ovaries) allow for animal-based seed dispersal and deposition with plenty of fertilizer. Flowers evolved to facilitate animal, especially insect, based pollen dispersal. Petals are modified leaves. Angiosperms currently dominate the flora of the world — over three fourths of all living plants are angiosperms.

Insects evolved from primitive segmented arthropods. The mouth parts of insects are modified legs. Insects are closely related to annelids. Insects dominate the fauna of the world. Over half of all named species are insects. One third of this number are beetles.

The end of the Cretaceous, about 65 million years ago, is marked by a minor mass extinction. This extinction marked the demise of all the lineages of dinosaurs save the birds. Up to this point mammals were confined to nocturnal, insectivorous niches. Once the dinosaurs were out of the picture, they diversified. *Morgonucudon* , a contemporary of dinosaurs, is an example of one of the first mammals. Mammals evolved from therapsid reptiles. The finback reptile *Diametrodon* is an example of a therapsid. One of the most successful lineages of mammals is, of course, humans. Humans are neotenous apes. Neoteny is a process which leads to an organism reaching reproductive capacity in its juvenile form. The primary line of evidence for this is the similarities between young apes and adult humans. Louis Bolk compiled a list of 25 features shared between adult humans and juvenile apes, including facial morphology, high relative brain weight, absence of brow ridges and cranial crests.

The earth has been in a state of flux for 4 billion years. Across this time, the abundance of different lineages varies wildly. New lineages evolve and radiate out across the face of the planet, pushing older lineages to extinction, or relictual existence in protected refugia or suitable microhabitats. Organisms modify their environments. This can be disastrous, as in the case of the oxygen holocaust. However, environmental modification can be the impetus for further evolutionary change. Overall, diversity has increased since the beginning of life. This increase is, however, interrupted numerous times by mass extinctions. Diversity appears to have hit an all-time high just prior to the appearance of humans. As the human population has increased, biological diversity has decreased at an ever-increasing pace. The correlation is probably causal.

### *Scientific Standing of Evolution and its Critics*

The theory of evolution and common descent were once controversial in scientific circles. This is no longer the case. Debates continue about how various aspects of evolution work. For example, all the details of patterns of relationships are not fully worked out. However, evolution and common descent are considered fact by the scientific community.

Scientific creationism is 100% crap. So-called "scientific" creationists do not base their objections on scientific reasoning or data. Their ideas are based on religious dogma, and their approach is simply to attack evolution. The types of arguments they use fall into several categories: distortions of scientific principles ( the second law of thermodynamics argument), straw man versions of evolution (the "too improbable to evolve by chance" argument), dishonest selective use of data (the declining speed of light argument) appeals to emotion or wishful thinking ("I don't want to be related to an ape"), appeals to personal incredulity ("I don't see how this could have evolved"), dishonestly quoting scientists out of context (Darwin's comments on the evolution of the eye) and simply fabricating data to suit their arguments (Gish's "bullfrog proteins").

Most importantly, scientific creationists do not have a testable, scientific theory to replace evolution with. Even if evolution turned out to be wrong, it would simply be replaced by another scientific theory. Creationists do not conduct scientific experiments, nor do they seek publication in peer-reviewed scientific journals. Much of their output is "preaching to the choir."

The most persuasive creationist argument is a non-scientific one — the appeal to fair play. "Shouldn't we present both sides of the argument?," they ask. The answer is no — the fair thing to do is exclude scientific creationism from public school science courses. Scientists have studied and tested evolution for 150 years. There is voluminous evidence for it. Within the scientific community, there are no competing theories. Until scientific creationists formulate a scientific theory, and submit it for testing, they have no right to demand equal time in science class to present their ideas. Evolution has earned a place in the science curriculum. Creationism has not.

Science is based on an open and honest look at the data. Much of creationism is built on dishonest debating techniques and special pleading for a case the data does not support. Science belongs in science classes. Evolution is science. Creationism is not. It's that simple.

The creationist attack on public school education means that school children are denied the possibility of learning about the most powerful and elegant theory in biology. Politicians are willing to allow the scientifically ignorant, but politically strong, to wreck the educational system in exchange for votes. People interested in evolution, and science education in general, need to closely watch school board elections. Creationist "stealth" candidates have been elected in several regions. Thankfully, many have been voted out once their views became apparent.

The majority of Americans are religious, but only a minority are religious nuts. The version of religion the far right wants to impose on America is as repulsive to most mainstream Christians as it is to members of other religions, atheists and agnostics. Most informed religious people see no reason for biological facts and theories to interfere with their religious beliefs.

### *The Importance of Evolution in Biology*

"Nothing in biology makes sense except in the light of evolution." — Theodosius Dobzhansky

Evolution has been called the cornerstone of biology, and for good reasons. It is possible to do research in biology with little or no knowledge of evolution. Most biologists do. But, without evolution biology becomes a disparate set of fields. Evolutionary explanations pervade all fields in biology and brings them together under one theoretical umbrella. We know from microevolutionary theory that natural selection should optimize the existing genetic variation in a population to maximize reproductive success. This provides a framework for interpreting a variety of biological traits and their relative importance. For example, a signal intended to attract a mate could be intercepted by predators. Natural selection has caused a trade- off between attracting mates and getting preyed upon. If you assume something other than reproductive success is optimized, many things in biology would make little sense. Without the theory of evolution, life history strategies would be poorly understood.

Macroevolutionary theory also helps explain many things about how living things work. Organisms are modified over time by cumulative natural selection. The numerous examples of jury- rigged design in nature are a direct result of this. The distribution of genetically based traits across groups is explained by splitting of lineages and the continued production of new traits by mutation. The traits are restricted to the lineages they arise in.

Details of the past also hold explanatory power in biology. Plants obtain their carbon by joining carbon dioxide gas to an organic molecule within their cells. This is called carbon fixation. The enzyme that fixes carbon is RuBP carboxlyase. Plants using C3 photosynthesis lose 1/3 to 1/2 of the carbon dioxide they originally fix. RuBP carboxlyase works well in the absence of oxygen, but poorly in its presence. This is because photosynthesis evolved when there was little gaseous oxygen present. Later, when oxygen became more abundant, the efficiency of photosynthesis decreased. Photosynthetic organisms compensated by making more of the enzyme. RuBP carboxylase is the most abundant protein on the planet partially because it is one of the least efficient.

Ecosystems, species, organisms and their genes all have long histories. A complete explanation of any biological trait must have two components. First, a proximal explanation — how does it work? And second, an ultimate explanation — what was it modified from? For centuries humans have asked, "Why are we here?" The answer to that question lies outside the realm of science. Biologists, however, can provide an elegant answer to the question, "How did we get here?"

### Modern Evolutionary Synthesis

The modern evolutionary synthesis (often referred to simply as the new synthesis, the modern synthesis, the evolutionary synthesis, neo-Darwinian synthesis or neo-Darwinism), generally denotes the integration of Charles Darwin's theory of the evolution of species by natural selection, Gregor Mendel's theory of genetics as the basis for biological inheritance, random genetic mutation as the source of variation, and mathematical population genetics. Major figures in the development of the modern synthesis include Thomas Hunt Morgan, R. A. Fisher, Theodosius Dobzhansky, J.B.S. Haldane, Sewall Wright, Julian Huxley, Ernst Mayr, George Gaylord Simpson, and G. Ledyard Stebbins.

Essentially, the modern synthesis introduced the connection between two important discoveries; the units of evolution (genes) with the mechanism of evolution (selection). It also represents a unification of several branches of biology that previously had little in common, particularly genetics, cytology, systematics, botany, and paleontology.

#### *History*

George John Romanes coined the term *neo-Darwinism* to refer to the theory of evolution preferred by Alfred Russel Wallace *et al.* Wallace rejected the Lamarckian idea of inheritance of acquired characteristics, something that Darwin, Huxley *et al* wouldn't rule out. The most prominent "neo-Darwinian" of the time after Darwin was August Weismann, who argued that hereditary material, which he called the germ plasm, was kept utterly separate from the development of the organism. This was seen by most biologists as an extreme position, however, and variations of neo-Lamarckism, orthogenesis ("progressive" evolution), and saltationism (evolution by "jumps" or mutations) were discussed as alternatives.

In 1900, Mendelian inheritance was "rediscovered", and was initially seen as supporting a form of "jumping" evolution. The biometric school, led by Karl Pearson and Walter Frank Raphael Weldon, argued against it vigorously, stating empirical evidence indicated that variation was continuous in most organisms. The Mendelian school, led by William Bateson, countered that in some cases the Mendelian evidence was indisputable and that future work would reveal its larger truth. Mendelism was taken up by many biologists, even though it was still extremely crude at this early stage. Its relevance to evolution was still hotly debated.

A critical link between experimental biology and evolution, as well as between Mendelian genetics, natural selection, and the chromosome theory of inheritance, arose from T. H. Morgan's work with the fruit fly *Drosophila melanogaster*. In 1910, Morgan discovered a mutant fly with solid white eyes (wild-type *Drosophila* have red eyes), and found that this condition—though appearing only in males—was inherited precisely as a Mendelian recessive trait. In the subsequent years, he and his colleagues developed the Mendelian-Chromosome theory of inheritance and Morgan and his colleagues published *The Mechanism of Mendelian Inheritance* in 1915. By that time, most biologists accepted that genes situated linearly on chromosomes were the primary mechanism of inheritance, although how this could be compatible with natural selection and gradual evolution remained unclear. Morgan's work was so popular that it is considered a hallmark of classical genetics.

This issue was partially resolved by R. A. Fisher, who in 1918 produced a paper entitled *The Correlation Between Relatives on the Supposition of Mendelian Inheritance*,[1] which showed using a model how continuous variation could be the result of the action of many discrete loci. This is sometimes regarded as the starting point of the synthesis, as Fisher was able to provide a rigorous statistical model for Mendelian inheritance, satisfying both the needs (and methods) of the biometric and Mendelian schools.

Morgan's student Theodosius Dobzhansky was the first to apply Morgan's chromosome theory and the mathematics of population genetics to natural populations of organisms, in particular *Drosophila melanogaster*. His 1937 work *Genetics and the Origin of Species* is usually considered the first mature work of neo-Darwinism. This work, as well as works by Ernst Mayr (*Systematics and the Origin of Species* – systematics), G. G. Simpson (*Tempo and Mode in Evolution* – paleontology) , and G. Ledyard Stebbins (*Variation and Evolution in Plants* – botany), are considered the four canonical works of the modern synthesis. C. D. Darlington (cytology) and Julian Huxley also wrote on the topic; Huxley coined both evolutionary synthesis and modern synthesis in his semi-popular work *Evolution: The Modern Synthesis* in 1942.

### *Tenets of the Modern Synthesis*

According to the modern synthesis as established in the 1930s and 1940s, genetic variation in populations arises by chance through mutation (this is now known to be sometimes caused by mistakes in DNA replication) and recombination (crossing over of homologous chromosomes during meiosis). Evolution consists primarily of changes in the frequencies of alleles between one generation and another as a result of genetic drift, gene flow and natural selection. Speciation occurs gradually when populations are reproductively isolated, e.g. by geographic barriers.

### *Further Advances*

The modern evolutionary synthesis continued to be developed and refined after the initial establishment in the 1930s and 1940s. The work of W. D. Hamilton, George C. Williams, John Maynard Smith and others led to the development of a gene-centric view of evolution in the 1960s. The synthesis as it exists now has extended the scope of the Darwinian idea of natural selection, specifically to include subsequent scientific discoveries and concepts unknown to Darwin such as DNA and genetics that allow rigorous, in many cases mathematical, analyses of phenomena such as kin selection, altruism, and speciation.

A particular interpretation of neo-Darwinism most commonly associated with Richard Dawkins asserts that the gene is the only true unit of selection. Dawkins further extended the Darwinian idea to include non-biological systems exhibiting the same type of selective behavior of the 'fittest' such as memes in culture.

"Increasingly, studies of genes and genomes are indicating that considerable horizontal transfer has occurred between prokaryotes." Horizontal gene transfer is called by some "A New Paradigm for Biology" and emphasised by others as an important factor in "The Hidden Hazards of Genetic Engineering". "While horizontal gene transfer is well-known among bacteria, it is only within the past 10 years that its occurrence has become recognized among higher plants and animals. The scope for horizontal gene tıansfer is essentially the entire biosphere, with bacteria and viruses serving both as intermediaries for gene trafficking and as reservoirs for gene multiplication and recombination (the process of making new combinations of genetic material)." [4] This approach is taken to its logical extreme by Lynn Margulis in her theory of symbiogenesis, with symbiosis as the major source of inherited variation, in which entire genomes are combined.

## How are Genes Engineered

A gene consists at an average of about 3000 "code syllables". So it is a fairly large piece of information that is to be transferred.

1. The desired gene is taken out of the DNA of the donor cell. Special techniques are used, that are of no imporance here.
2. A gene is much too small to be inserted with some kind of microsurgery. Therefore some carrier ("vector") is required which takes the gene into the recipient cell and somemhow gets it inserted into the DNA of the recipient.

### *Methods for Gene Insertion*

Different means are used for carrying the desired gene into the hereditary substance.. For plants, the most common method is the use of a bacterium, Agrobacterium Tumefacsiens. This method is however not usable for cereals. For those, insertion with the "gene cannon" or microinjection is used. For animals, certain viruses are used. These methods will be described briefly below.

#### *1. Bacteria as gene carriers*

A bacteria with the name Agrobacterium Tumefasciens has the ability to induce a kind of benign tumor growth in the plant it infects. This it achieves by inserting some genes into the DNA of the plant. In genetic engineering, the gene that is to be inserted is hooked on to the bacterial

DNA. Then the recipient cells are exposed to the bacterium, so that they become infected by it. The bacterial DNA with the engineered gene is then inserted into the DNA of the recipient.

*2. Viruses as gene carriers*

Viruses are pieces of DNA encapsulated within a protein shell. Their DNA enters the cell at infection and goes to the cell nucleus. There the virus DNA forces the cell to make many thousands of copies of the virus. The viruses used for gene transfer are modified so that they will not induce virus copying in the cell. They will introduce the desired DNA into the DNA of the recipient in similar manner as the Bacterium tumefasciens. Along with it some virus genes also are inserted.

*3. Mechanical methods*

*The gene cannon:* It uses very small golden beads as cannon balls. The desired gene is "glued" on to the surface of the ball. Thereafter the ball is shot into the cell nucleus.

*Microinjection:* A solution with the desired gene is injected into the cell.

*Tedious and expensive:* The insertion methods have all in common that it is impossible to control where the gene will attach . It is a completely random procedure. So the inserted DNA can be hooked on to the recipient DNA anywhere. It is a sheer matter of luck if it happens to stick in the very small proportion of DNA that is active. And even then, there is a risk that the insertion will have occurred in such a place that the result is a disfigured, weak or disease-prone organism (which has been repeatedly observed to be the case).

Therefore genetic engineering is a tedious and very costly procedure. Only a small proportion of all the attempts will result in an organism that seems "substantially equivalent" with the natural counterpart. But even then, it may have a chemical abnormality that is hazardous to the health of those eating it. The trouble is that there is no fully reliable method to detect dangerous substances. And the existing methods for safety testing are very costly and time consuming.

This is why the biotechnology industry has taken great pains to make the regulators to regard genetic engineering as just a variation of breeding. If so, they can accuse the authorities for unjust discrimination of their products. - If food from conventionally bred organisms must not be tested, why should food from organisms bred through genetic engineering have to be tested is their argument. Unfortunately, in the first round, the industry has been successful with this strategy. They have been able to make the regulators believe that there is no important difference between genetic engineering and conventional breeding. This claim has however no scientific basis, see *"Conventional breeding is fundamentally different from genetic engineering". And most importantly, the differences are so important that consumers will be exposed to potentially serious health risks if they are neglected.*

*The problematic "insertion package":* The proponents of biotechnology seldom mention that the desired gene is very seldom inserted alone. To make the insertion successful and effective, other genes, taken from various organisms, including viruses, plants or bacteria need to be added not. In addition to the carrier genes mentioned above there are two other important categories of genes that commonly are added in the "insertion package". These are presently marker genes, promoter genes and species barrier penetration genes.

*Markers:* Even if the gene package has become

successfully inserted, it may attach to a region of non-active DNA. Then the gene marker will be silenced and the cell will not survive antibiotics. So only genes that have been inserted into active regions, will express antibiotics resistance as well as the properties conferred by the Trait Gene. There is no way of seeing if the insertion has been successful. Therefore some “marker gene” is used, that is inserted along with the desired gene.

The success of gene insertion is tested by adding the antibiotic to the culture of recipient cells. The idea is to find out which ones are resistant to the antibiotic. So the cells that survive the exposure to the antibiotic are those that carry a successfully inserted resistance gene. Then, most probably, the desired gene also has been inserted. The most common marker so far has been a gene that confers resistance to Kanamycin, an antibiotic that belongs to a group of valuable antibiotics. Even if it is itself not used much today, it is closely related to valuable antibiotics. And there is a tendenency for cross resistance. That is resistance to Kanamycin may also confer resistance to its relatives. Another marker, used a/o. in Novartis Basta Corn is a gene that confers resistance to Ampicillin, a valuable and commonly used antibiotic. It is suspected that the antibiotic gene markers might lead to an increase of the numbers of bacteria that are resistant to the antibiotic.

*Promoters:* Even when the gene package has been inserted in an active region, the Trait Gene may express the desired trait only weakly or to a varying degree (that is, the production of the corresponding protein is not stable and high).Therefore, to ensure that the Trait Gene will express the trait persistently, a promoter gene is added to the insertion package. Promoter genes are natural parts of chromosomes. Their function is to enhance the activity of a certain gene. Normally these are under the control of regulator genes that turn them on and off. But in genetic engineering the corresponding regulators are normally not present as the promoter comes from an other species. This means that the promoter will exert a strong and persistent stimulating influence on the Trait Gene. This is basically unnatural as normally the activities of all genes are regulated in response to the conditions in the cell. It means that the fine tuned cellular feed-back balance system is persistently disturbed in this respect. The promoter gene commonly used in genetically engineered plants is the Cauliflower Mosaic Virus (CaMV) promoter. It is a strong promoter that has a high species compatibility. A potential problem however is that this virus gene may combine with infecting viruses so as to give rise to new viruses.

*Barrier penetration genes:* The cell has the ability to identify genes that are alien to the species. There are strong mechanisms that prevent such alien genes from combining with the genome. They break down foreign genes, or prevent their replication, or cut out and inactivate a foreign gene that has been able to insert itself. This is called the "species barrier".

This barrier is a major problem in genetic engineering. To overcome it, gene technology has invented various combinations of genes that have the ability to promote penetration of the species barrier. They consist of genes from viruses or bacteria that have the faclitate the insertion of genes into a new host.

An important complication is that these genes may spread in nature and contribute to increased species-barrier transfers. This is suspected to cause the generation of new and hazardous bacteria. Another complication is that the inserted virus genes may recombine with infecting viruses and give rise to new viruses and thereby new diseases.

*Unpredictable*

The randomness of insertion makes it impossible to predicts it's effects. But even when the location of the inserted gene is exactly known, the knowledge of Molecular Biology is far too incomplete to make it possible to predict the effects of the insertion. Actually genetic engineering is based on an outdated theory about genes which underestimated the consequences of the insertion of a gene into a completely new environment. It has turned out that there is a much greater interdependence between genes than formerly believed. This makes it impossible to consider a gene a carrier of a specific trait. In reality, the same gene may have different effects in different environments, see "The outdated basis of genetic engineering". In addition to this, there are additonal factors that contribute to unpredictablility. One is that the promoter gene does not only influence the activity of the inserted gene but may also promote the activity of adjacent genes of the recipient. This may give rise to important metabolic imbalances that may generate unexpected harmful substances. Also the insertion of a foreign sequence of genetic instructions in the finely attuned seuqence of genetic instructions of the recipient organism may disrupt the close control over metabolic processes exerted by DNA. This may also generate unexpected harmful substances.

Gene technology is still in a primitive stage. The insertion of genes is haphazard and therefore yields unpredictable results. Moreover, the knowledge of molecular biology is much too incomplete to be able to predict the effects of an inserted gene even when the position of its insertion is exactly known. To make the insertion of a desired gene successful, other genes have to be added that add to the unpredictability of gene insertion. Moreover these added genes in the "insertion package" may give rise to potentially serious problems with new viruses and bacteria. Much too little is known today about these problems to justify a release into nature of GE organsims. So genetic engineering is not just the simple addition of some "desired trait" as the biotechnology proponents have been putting it. It is the haphazard insertion of a set of foreign genes or gene fragments out of which one is the "desired trait" gene. This insertion has unpredictable effects within the organism as well as potentially problematic effects on their new environment.

**Sequencing of Entire Human Genome**

Human genome refers to effort of sequencing and mapping of the entire human DNA. This work is being done two ways: firstly by *HGP - Human Genome Project* (see http://www.nhgri.nih.gov/HGP/ ) - an international and inter-institutional research effort; and secondly by several private companies like *Celera* (see http://www.celera.com/ ) owned by Mr. Ventair. It is expected that the sequencing will be completed about year 2002. Mr. Ventair got from private investors over 300 millions of US dollars. He claims that his company *Celera* will get results earlier than HGP, but he, after completion of sequencing of the whole human genome, wishes to provide results to interested parties with delays and for money. At least he does not require from his customers, i.e. those who will buy access to his databases, that he will obtain copyrights of technologies developed on the basis of these databases - some other private companies require for example participation in property of patents developed with help of their databases! *Celera*'s technique relies on splitting of the whole human genome into pieces and then putting them together. *HGP* on the other hand is going through whole genome step by step forwards.

# 3

# Practices of Biotechnology: A Modern Perspective

## Exopolysaccharides Production in *Rhizobium* sp.

The combined effects of the processing parameters for exopolysaccharides production by *Rhizobium* sp. was studied using the experimental design and response surface methodology. The experiments were carried out using a fermenter with 20 L capacity, as the reactor. All processing parameters were online monitored. The temperature [(30 ± 1)°C] and pH value (7.0 ± 0.1) were kept constant throughout the experimental time. As statistical tools, a complete $2^3$ factorial planning with central point and response surface were used to study the interactions among three relevant variables of the fermentation process: calcium carbonate concentration, aeration and agitation. The processing parameters setup for reaching a maximum response for exopolysaccharides production was obtained when applying the highest values for calcium carbonate concentration (1.1 g/L), aeration (1.3 vvm) and agitation (800 rpm). In addition, the combination of these optimum processing parameters yielded $Y_{P/S}$ (g/g) = 0.35. Bacterial exopolysaccharides are extensively used as thickening and gelling agents in a wide range of industrial products and processes due to their structural and physical properties diversity (Copetti et al. 1997; Rinaudo, 2001). The structure, composition and viscosity of the microbial polysaccharides depend on several factors, such as the composition of the culture medium, carbon and nitrogen source, mineral salts, trace elements, type of strain, and fermentation conditions (pH, temperature, oxygen concentration, agitation) (Moreira et al. 1998; Weuster-Botz, 2000; Pinto et al. 2002; Duta et al. 2004). For many microbial species, calcium is essentially required in small amounts: it is essential in maintaining cell wall rigidity, it stabilizes oligomeric proteins and covalently bound protein peptidoglycan complexes in the outer membrane, as well as have a requirement for chemotaxis (Macció et al. 2002). The reach of optimized fermentation conditions, particularly associated to physical and chemical parameters, is of primary and great importance for the development of any process, due to their impact upon its economics and practicability. The diversity of combinatory interactions among medium components, metabolism of cells and the large number of chemical requirements for processing metabolic products, do not allow satisfactory detailed modelling. The one-dimensional search with successive changes on variables conditions is still employed, even trough it is well accepted that it is practically impossible for the one-dimensional search to accomplish an appropriate optimum combination in a finite number of experiments. Single variable optimization methods are not only tedious, but can also lead to misinterpretation of results, especially taking into account that the interaction between different factors is overlooked (Abdel-Fattah et al. 2005). Statistical experimental designs have been used for many decades and can be adopted on several steps of an optimization strategy, such as for screening experiments or searching for the optimal conditions of a targeted response (Kim et al. 2005; Lee and Gilmore, 2005; Nawani and Kapadnis, 2005; Senthilkumar et al. 2005; Wang and Lu, 2005). Recently, the results analyzed by a statistical planned experiment are better acknowledged than those carried out by the traditional one-variable-at-a-time method. Some of the popular choices, applying statistical designs to bioprocessing, include the Plackett-Burman design (Liu et al. 2003; Wang and Lu, 2005) and response surface methodology with various designs (Abdel-Fattah, 2002; Abdel-Fattah and Olama, 2002; Tanyildizi et al. 2005). The response surface methodology is an empirical modelling system that assesses the relationship between a group of variables, which can be controlled experimentally, and the observed response. This methodology is applied mainly both in food science and in the optimization of fermentative processes. It is a 2-level factorial design, where contour plots are generated by linear or quadratic effects of key variables, and a model equation is derived, fitting the experimental data to the calculate system's optimal response (Lakshman et al. 2004; Cazetta et al. 2005; Khanna and Srivastava, 2005). Fast growing rhizobia synthesize different extracellular polysaccharides, like acid exopolysaccharides (EPS) of high molecular weight (Zevenhuizen, 1986). Those microorganisms can also produce neutral glucans, formed by â-1,4 bonds, which are found as cellulosic microfibrils of low molecular weight (Zevenhuizen, 1986; Breedveld et al. 1990; Jain et al. 1990; Breedveld et al. 1993). Many rhizobia stains produced polysaccharides are not found freely dispersed in the medium, but attached upon the microbial cell as an amorphous viscous material. Among these, the curdlana homo-polymer is outstanding, in which the derived sulphated sites show anticoagulant and anti-thrombosisactivities. Besides, they also present an inhibiting effect against the HIV-1 virus *in vitro* infection (Jagodzinski et al. 1994). Curdlana sulpho-alkiled sites present anti-tumour activity, inhibiting the development of the Sarcoma tumours 180 (Bohn, 1995). The present study investigated the relationship among three variables, that is, calcium carbonate concentration, agitation, and aeration, in exopolysaccharide production by *Rhizobium* sp.

**Microorganism**

The *Rhizobium* EQ1 strain was used in the experiments. This was isolated from a variety of beans culture, native from an arid region of the Northeast of Brazil, named as "Caupi". This strain has been kept in yeast mannitol agar (YMA), for several months, being catalogued, after its characterization, at the culture bank of the Biochemical Engineering Department from Federal University of Rio de Janeiro, under the code EQ1. The microorganism was grown under (30 ± 1)°C for 48 hrs, using laboratory tubes filled with YMA medium. After growth, the cultures were stored at (5 ± 1)°C.

**Culture Media**

The microorganism stock culture was maintained in modified yeast mannitol agar extract, which presented the following composition (g/L): mannitol (10.0); $K_2HPO_4$ (0.1); $KH_2PO_4$ (0.4); $MgSO_4.7H_2O$ (0.2); NaCl (0.1); yeast extract (0.4); agar (15.0). The medium pH was adjusted to 7.0 (Jordan, 1984). The stock culture was used for preparing the inoculum, using 500 mL Erlenmeyer flasks, containing 100 mL of the YMA medium free from agar. Incubation was carried out in a rotary shaker at 200 rpm and (30 ± 1)°C, for 48 hrs, with the cells in the final of exponential growth phase. The exopolysaccharides production assays were carried out using a similar culture medium, where the YMA medium was: supplemented with manganese ions ($MnCl_2$. $4H_2O$ 0.12 g/L); free from agar; and the calcium carbonate concentration ranged in accordance to the experimental design ($CaCO_3$ 0.5-1.0 g/L). The pH of this medium was adjusted to 7.0 by the addition of NaOH (50% w/v). All culture media were sterilized at 121°C for 20 min. The inoculation of the production medium was made in order to obtain an average cells concentration of 0,77 ± 0,02 mg/mL.

**Production of Exopolysaccharides in Bioreactors**

These experiments were conducted in a fermenter (Model BioFlow IV, New Brunswick Scientific), with 20 L capacity, equipped with disc impeller, oxygen and pH electrodes. The equipment also monitored temperature, agitation speed, gas purging flow rate, pumping rates, antifoam addition and the vessel level. All processing parameters were online monitored, with the aid of AFS 3.0 software (Advanced Fermentation Software, New Brunswick Scientific). The temperature (30 ± 1°C) and pH value (7.0 ± 0.1) were kept constant during the experiments. Other parameters, like substrate concentration, aeration and agitation, were chosen as the most significant ones, considering the experimental design. After selecting those parameters, experiments were done in duplicate, for superior (+) and lower (-) levels of the experimental design, and in triplicate, for the central point (0). For each experiment, 1000 mL of the inoculum was used, that is, 10% (v/v) of the initial working volume (10 L). The process was conducted throughout 48 hrs.

**Analytical Methods**

During the process, microscopic examinations, using Gram method, were performed in order to detect possible microbial contaminations in the medium. Prior for the quantitative determination of mannitol, the fermented broth was filtered through 0.2 $\mu$m Millipore membranes, in order to remove microbial cells. In the filtered fluid, the substrate was analyzed by high performance liquid chromatography (HPLC), in a Waters chromatograph, equipped with SHODEX SC1011 ion-exchange columns, at 75°C. Reagent water type I (ASTM, 2001) was used as eluent, and the elution rate applied was 0.8 mL/min. The amount of fermented exopolysaccharide was determined by dry-weight measurements. The fermented broth was heated at (80 ± 1)°C, for 10 min, to ensure microbial inactivation. Afterwards, the microbial cells were removed by a filtration step. In order to precipitate the exopolysaccharides, a solution of ethanol P.A. and reagent water type I (ASTM, 2001) (3:1) was added to the fermented broth. After the exopolysaccharides total precipitation, the suspended material was filtered through 0.2 $\mu$m Millipore membrane, using Gouche crucible previously weighed. The obtained product was dried at (80 ± 1)°C until constant weight. All determinations were done in triplicate. The exopolysaccharides extracted from the fermented broth was purified through successive washings with solutions of ethanol P.A. and reagent water type I (ASTM, 2001) at 70, 80 and 90% (v/v), respectively. The product was finally dried by a nitrogen gas purging flow, under controlled heating.

### Experimental Design and Statistical Analysis

This statistical technique is widely used as a tool to verify the efficacy of several processes. In the present work, it has been used for obtaining pieces of information about the exopolysaccharides production process; thus, a reduction in operational costs can be expected. A $2^3$ factorial planning, in duplicate, with central point in triplicate, was used (Box et al. 1978; Neto et al. 1995). Additionally, each experiment was repeated once, around the central point neighbourhood, while the central point was repeated twice, leading to a set of 19 experiments. Three central points were added to estimate the experimental error and to investigate the suitability of the proposed model. The independent variables studied, $X_1$ calcium carbonate concentration (g/L), $X_2$ aeration (vvm), $X_3$ agitation (rpm). The manipulation responses of the input variables were evaluated as a function of the substrate conversion into exopolysaccharide, coded by $Y_{p/s}$ (g/g). A mathematical model, describing the relationships among the process dependent variable, Yp/s, and the independent variables in a second-order, was developed.

Design-based experimental data were matched according to the following second-order polynomial. All the calculations involved as well as the drawing of all three-dimensional surface (3D) have been obtained using the Statistica™ Software for Windows, Version 5.5 computer package, produced by Stat Soft. The model allowed the evaluation of the effects of linear, quadratic and combined effects of the independents variables upon the variable dependent variable. The Student's *t*-test was employed in order to check the statistical significance of the regression coefficients. The Fisher's F-test for analysis of variance (ANOVA) was performed on experimental data to evaluate the statistical significance of the model. Three-dimensional surface (3D) plots were drawn to illustrate the main and interactive effects of the independent variables on exopolysaccharides production. The optimum values of the selected variables were obtained both by solving the regression and also by analyzing the response surface contour plots (Myers and Montgomery, 2002).

The statistical technique is widely used as a tool for checking the efficiency of several processes. In the present work it has been used with the purpose of obtaining information about the exopolysaccharides production process; consequently, a reduction in the operational variability and a cut down in operational costs can be expected. The experimental results ($Y_{p/s}$), associated to the processing set up of each independent variables. Using the designed experimental data presented, the polynomial proposed model for $Y_{p/s}$ was regressed by only considering the significant terms. The expanded equation is shown below:

$$Y_{p/s} = -0.361612 + (1.147835\, X_1) + (-0.714687\ ) + (-0.081844\, X_2) + (-0.000081\, X_3) + (0.205141\, X_1X_2) + (0.000153\, X_1X_3) + (0.000052\, X_2X_3)$$

Besides the linear effect of the substrate/

exopolysaccharides factor, $Y_{p/s}$, the response surface method also gives an insight about the parameters quadratic and combined effects. These analyses were done by using both Fisher's F- test and Student *t*-test statistical tools. The student *t*-test was used to determine the significance of the parameters regression coefficients. The p-values were used as a tool to check the significance of the interaction effects, which in turn may indicate the patterns of the interactions among the variables. In general, larger magnitudes of *t* and smaller of p, indicates that the corresponding coefficient term (Myers and Montgomery, 2002). The regression coefficient, *t* and p values for all the linear, quadratic and combined effects, with a 95% significance level. It was observed that the coefficients for the linear and quadratic effects of the factor calcium carbonate concentration parameter, the linear effect of the aeration parameter, the combined effects between calcium carbonate concentration with the aeration and agitation parameters ($p = 0.000$ for all) were highly significant.

The value observed for the factor agitation was slightly less significant ($p = 0.014$). The statistical significance of the ratio, between the of mean square variation, due to regression, and the mean square residual error, was tested using analysis of variance (ANOVA). ANOVA is a statistical technique that subdivides the total variation of a set of data into component associated to specific sources of variation for the purpose of testing hypotheses for the modelled parameters. According to the ANOVA, the F-values for all regressions were high, what indicates that most of the variations on the response variable can be explained by the regression equation. The associated p-value is used to estimate whether F is large enough to indicate statistical significance. A p-value lower than 0.01 indicates that the model is considered to be statistically significant (Kim et al. 2003). The p values of all of the regression were lower than 0.01. This means that at least one of the terms in the regression equation has a significant correlation with the response variable. The ANOVA, which measures the amount of variation in the response data left unexplained by the model. The type of the model, chosen to explain the relationship between the factors and response, is correct. The analysis of variance (ANOVA), indicates that the second-order polynomial model was highly significant and adequate to represent the actual relationship between the response and input variables, with very small p values ($p = 0.0000$). All the linear terms of the model were significant for the set confidence level, as well as the quadratic term of $X_1$ variable.

It can also be seen that the interaction term among the three variables ($X_1X_2X_3$), was not statistically significant. Statistical data analysis shows that calcium carbonateconcentration is the most important variable for the production process. The matching quality, of the data obtained by the model proposed, was evaluated considering the correlation coefficient, $R^2$, between the experimental and modelled data. The mathematical adjust of those values generated a $R^2 = 0.9884$, revealing that the model could not explain only 1.16% of the overall effects, showing that it is a robust statistical model. The regression plot of the $Y_{p/s}$ experimental values against those predicted, revealing a linear mathematical relation among them. In addition, the mismatching analysis and further error terms, agree the adequacy of the predicted data, thus the reliability of the model proposed. The 3D response surface plots described by the regression model were drawn to illustrate the effects of the independent variables, and combined effects of each independent variable upon the response variable. 3D response surface based on the $Y_{p/s}$ response against the variation of calcium carbonate concentration ($X_1$) and aeration ($X_2$) independent variables upon $Y_{p/s}$, while the third independent variable, agitation ($X_3$), was kept constant level (800 rpm). It can be observed that the maximum estimated $Y_{p/s}$ 0.3431 (g/g) was obtained using calcium carbonate concentration of 1.0 g/L and aeration of 1.3 vvm. The data obtained by varying calcium carbonate concentration ($X_1$) and agitation ($X_3$), fixing aeration at 1.3 vvm, can be observed. The analysis reveals that the maximum substrate conversion into exopolysaccharides was also obtained under the following condition: at calcium carbonate concentration of 1.0 g/L and agitation of 800 rpm. An increase on agitation and aeration parameters promotes an increase on $Y_{p/s}$. For each combination of agitation and aeration values, the model generates a maximum value for the concentration of calcium carbonate. Thus, for the maximum values used in the experimental design, 1.3 vvm for the aeration and 800 rpm for agitation, the predicted value for of calcium carbonate concentration is 1.1 g/L. The application of the regression model for the substrate/exopolysaccharides factor, $Y_{p/s}$, was tested using calcium carbonate concentration of 1.1 g/L, agitation of 800 rpm and aeration of 1.3 vvm, with triplicate experiments. For this experimental condition, $Y_{p/s}$ mean value was of 0.3574 ± 0.008448 (g/g), which agrees with the predicted value, 0.3471 (g/g). This verification revealed a high accuracy of the model, that is, 97.12%, which is an evidence of the model validation, under the investigated conditions.

The process for exopolysaccharides production is carried out under aeration and agitation. The control of such parameters is of great importance for adequately conducting the fermentation process. According to Brock and Madigan (1991), for a rise in biomass from aerobic microorganisms, a vigorous aeration is required, what should be reached by forced aeration. This induced aeration is essential for getting high performance responses form the process, since oxygen is slightly soluble in water, not being quickly replaced by air diffusion, and worthwhile for microbial growth. Zevenhuizen (1986), using a mannitol-rich culture medium, has directed the polysaccharide synthesis towards exopolysaccharides by applying forced aeration. The efficiency in conducing forced aeration is linked to agitation, which favours oxygen diffusion in the medium and its transfer to cells. Agitation also promotes a reduction in nutrient particles, favouring the nutrient homogenization in the culture medium, providing additionally a rise in mass transfer rates, this favouring microbial growth. In the production medium used for obtaining biopolymers, several ions are added to propitiate the exopolysaccharides, and these shall be in appropriate amounts. The metallic ions perform catalytic and essential structural functions in proteins, being accumulated inside the cell by active transport (Macció et al. 2002). The literature tells about the use of calcium carbonate to prevent the acidification of the bacterial broth (Macció et al. 2002). Jordan (1984) suggested the use of 4.0 g/L of calcium carbonate in the culture media, for controlling pH in the *Rhizobium* sp culture. O' Hara et al. (1989) previously reported that 1 and 2 mM of calcium was necessary for cytoplasmic pH maintenance in *Rhizobium meliloti* acid-sensitive strain.

The use of calcium ions can be intimately related to the stabilization of proteins involved in the exopolysaccharides synthesis process (Soto et al. 2004). In agreement with the data presented in this study the yield income values of the product ($Y_{p/s}$), in a medium containing 1.0 g/L of calcium carbonate was 0.35 g/g, in average. These results are consistent with the biochemical processes involving polymerization reactions forming high-viscosity products.

Most researches developed using *Rhizobium* genus bacteria are related to genetics and bacteria-host plant symbiotic interactions issues. Little is known about the production of extra-cellular polysaccharides by *Rhizobium*, as well as their properties in solution. In addition, no studies on monitoring medium composition, agitation and aeration parameters effects, on exopolysaccharides production, were found.

### Effect of Glutaraldehyde Biocide on Laboratory-Scale Rotating Biological Contactors and Biocide Efficacy

The effect of glutaraldehyde, a commercial biocide widely used in paper and pulp industry, on the performance of laboratory-scale rotating biological contactors (RBCs) as well as biocide efficacy was studied. Biofilms were established on the RBCs and then exposed to 0 - 180 ppm glutaraldehyde at a dilution rate of 1.60 $h^{-1}$. The results showed that the biofilms became acclimated to glutaraldehyde and eventually could degrade it. Acclimation to the biocide took longer at the higher biocide concentrations. The degree of biocide degradation and chemical oxygen demand (COD) removal depended on acclimation period, the presence of other organic matters and the amount of mineral salts available. Glutaraldehyde at up to 80 ppm had no effect on treatment efficiency and populations of biofilms and planktonic phase of the system whereas glutaraldehyde at 180 ppm caused a progressive decline in all measured values. However, no glutaraldehyde concentration used in the study was sufficiently high to kill microorganisms in the RBC system. The presence of biofilm provided additional resistance to glutaraldehyde to bacteria because the biocide had to penetrate through biofilm to reach bacteria. The increased resistance of bacteria to glutaraldehyde due to acclimation should be considered in biocide applications.

Glutaraldehyde (1,5-pentanedial) is a non-oxidizing biocide which achieves its biocidal activity by involving the cross-linking of the outer proteinaceous layers of the cell in such a way that cellular permeability is altered. The bacterial cell is unable to undertake most, if not all, of its essential functions. The ability of the outer covering of the cell to transport nutrients to the cell and to remove waste products from the cell is hindered and cell death results (Russell and Chopra, 1996; Simons et al. 2000). Glutaraldehyde is widely used as an antimicrobial agent in a variety of applications such as in cooling water systems, paper-pulp industry, oilfield operation, leather tanning industry, poultry industry, cosmetic field, microbiological field, food industry and medical area. The extensive use of this biocide is due to being non-corrosive to stainless steel, soft metals, rubber and glass (Banner, 1995; Herbert, 1995; Lutey, 1995; Walsh et al. 1999).

Microbial biofilms play an essential role in engineered processes especially in biological treatment and energy production (Flemming, 1993; Bryers, 2000). They have been reported to have higher resistant to toxic compounds than free cell systems (Sofer et al. 1990; Lazarova and Manem, 2000). This makes them very useful in treating wastewaters containing toxic chemicals. However, suspended growth systems have the benefit of inherent dilution due to larger aeration volume as compared to attached-growth systems, which have known to be more susceptible to shock loadings. Biofilms can cause many negative effects commonly termed biofouling in industrial water systems. Biofilm formation causes reductions in efficiency in industrial machinery, heat exchangers, vessels and filters (Melo and Bott, 1997; Bott, 1998; Das et al. 1998) and may cause acceleration of metallic corrosion (Cloete et al. 1998). Numerous methods have been used for minimizing adverse effects of biofilm formation. The most widely practiced approach to control biofouling is by way of chemical treatment using biocides (Cheung and Beech, 1996; Stewart et al. 2000). The widespread application of biocides in many industries to control growth of planktonic and / or sessile microorganisms in a biofilm can cause environmental and ecological problems because sooner or later the biocides will have to be discharged into the environment either natural water or municipal effluent treatment plants (Nishiyama et al. 1995; Wyndham and Kennedy, 1995; Johnston et al. 1998; Reuter, 1998; Bailey et al. 1999; Sarlin et al. 1999; Juergensen et al. 2000; Nishihara et al. 2000). Therefore, knowledge of the effects of biocides on wastewater treatment units and their efficacy against bacterial cells is important.

Rotating biological contactors (RBCs) are one of the standard methods of biological wastewater treatment in which degradation and bio-oxidation are predominantly achieved by a biofilm which is attached to the rotating discs (Arundel, 1995; Banerjee, 1997; Casey, 1997). The RBC process is recommended for wastewater treatment because of simplicity in construction and maintenance, stable operation and relatively low cost. Laboratory-scale RBC units may be used to predict the effect of the biocide on the operation of full scale RBC, the ability of full scale RBCs to degrade biocides and the ability of biocides to control and eliminate biofilms in continuous systems. In this study the effects of 50% glutaraldehyde, a commercial biocide widely used in paper and pulp industry, on biofilms in RBCs and on biofilm establishment in RBCs after acclimation have been investigated. The effects of the biocide on the reduction in chemical oxygen demand (COD) and glutaraldehyde concentration of the synthetic wastewater applied to the RBCs were determined. The development of resistant / acclimatized microflora to the biocide and the effects that such microorganisms would have on biocide degradation as a sole carbon source and RBC treatment efficiency were observed. Differences in susceptibility of planktonic and biofilm bacteria to glutaraldehyde were also compared.

### *Materials and Methods*

Single-stage laboratory-scale RBC units were constructed in groups of three, the three units being served by a common motor and drive shaft . Each unit consisted of an influent chamber, a disc chamber and an in-built settling tank. These compartments were separated by PVC walls, each containing a 20-mm diameter transfer hole near the base of the wall through which liquor and suspended biomass passed. In each unit the treated liquor from the disc chamber was passed to the settling tank, where the biomass was removed, before being discharged via an overflow tube. Settled biomass was removed from each settling tank via a low-level gate valve. The configuration of the single stage laboratory scale RBC units. The standard synthetic wastewater consisted of lab-lemco broth, 90 mg $L^{-1}$ which composed of 33.75 mg $L^{-1}$ lab-lemco powder (Oxoid) and 56.25 mg $L^{-1}$ bacteriological peptone (Oxoid); $NH_4Cl$, 54 mg $L^{-1}$; $K_2HPO_4$, 28 mg $L^{-1}$; NaCl, 7 mg $L^{-1}$; $CaCl_2.2H_2O$, 4 mg $L^{-1}$ and $MgSO_4.7H_2O$, 2 mg $L^{-1}$ (modified from Laopaiboon et al. 2002).

An inoculum of recycled sludge (0.5 litres) from the wastewater treatment system of Phoenix Pulp and Paper Public Company Ltd., Khon Kaen, Thailand, was added into the disc stage of the RBC units filled with the synthetic wastewater. The units were operated as a batch system for 24 hrs before the flow of the synthetic wastewater was introduced to the units at a rate of 2.5 L $h^{-1}$. Approximately two weeks the surface of the discs was covered by a thin homogeneous biofilm about 1-2 mm thick. A commercial biocide, glutaraldehyde 50% aqueous solution (Ondeo Nalco,

Thailand) was then added to the synthetic wastewater to give final concentrations of 0, 40, 80 and 180 ppm (active ingredients). MIC of glutaraldehyde on three types of inoculum was determined. The three inocula were 'TSA cells' (planktonic cells from the disc stage of the control RBC inoculated onto TSA slant at 37°C for 24 hrs), 'control planktonic cells' (planktonic cells from the disc stage of the control RBC) and 'acclimatized cells' (planktonic cells from the disc stage at steady state of the experiment in next section). The MIC was determined by inoculating one ml of the inoculum into a series of sterile vials containing 9 ml of the synthetic wastewater and the desired concentration of glutaraldehyde. The vials were incubated at 37°C for 48 hrs. The lowest concentration of bactericide showing absence of growth by total viable count (TVC) measurement was taken to be MIC. The planktonic microorganisms (0.5 litres) which had previously been subjected to glutaraldehyde at 180 ppm for 18 days were added to the disc stage of the two clean RBC units filled with the synthetic wastewater containing 180 ppm glutaraldehyde. The units were operated as a batch system for 24 hrs before the flow of the wastewater A (the wastewater containing the mineral salts and 180 ppm glutaraldehyde) or the wastewater B (the wastewater containing only 180 ppm glutaraldehyde) was introduced to the units at a rate of 2.5 L $h^{-1}$. Biofilm formation on the RBC discs was then observed. If a biofilm was able to re-establish, all measurements as described below were performed. At the end of the experiment, single colonies of bacterial cells in the biofilms were isolated on tryptone soya agar (TSA) plate by spread plate technique. The colonies were tentatively identified by the API 20 NE strip test (BioMérieux SA, France).

### *Sampling and Analysis*

*Chemical oxygen demand and glutaraldehyde*

*concentration.* Influent and effluent samples were filtered through a 47 mm diameter, 0.2 $\mu$m pore size, cellulose acetate membrane filter (Whatman) to remove bacterial cells. The COD of the samples was then determined by the modified closed reflux, titrimetric method (APHA AWWA and WPCE, 1995). Standard ferrous ammonium sulphate, 0.025 M, was used as a titrant. Glutaraldehyde concentrations were measured by colorimetric micromethod (Boratynski and Zal, 1990).

*Enumeration of viable populations of biofilm and planktonic bacteria.* A one-cm$^2$ disc was cut from the biofilm using a sharp, sterilized cork borer (1.13 cm diameter). The biofilm was removed using a sterile scalpel and placed in 9 ml quarter strength Ringer's solution. The sample was sonicated for 5 min 3 times and vortex mixed between sonication for 10 sec (Stoodley et al. 1999) and a (TVC) was then performed. The liquid in the disc chamber was also sampled and the TVC of the planktonic organisms was determined. TSA was employed for all TVC's throughout the experiments and all plates were incubated at 37°C for 48 hrs.

*Population changes of higher organisms in the biofilm.* Samples of the biofilm were mounted on a microscope slide. Population changes of protozoa and mesozoan in the samples were observed by visual microscopy at magnification of (x 400).

### Results and Discussion

The single-stage laboratory-scale RBCs proved to be a very efficient system to establish a biofilm under the operating conditions used. Twenty four hours as a batch system at the beginning of the operation was sufficient for the induction phase of biofilm formation-organic absorption following by the transport and subsequent attachment of microorganisms to the surface (Trulear and Characklis, 1982). Some biofilms detached from the discs and re-established onto the RBC discs resulting in relatively stable biofilm thickness in all treatments. The results obtained also suggest that the polyvinylchloride is a good material for supporting good growth of a mixed population biofilm. This was supported by Apilánez et al. (1998) and Kerr et al. (1999) who found that the rate of biofilm formation and total viable population of biofilm depend on substratum. The biofilms on the discs of the treated RBC units with glutaraldehyde were markedly thicker than those of the control RBC unit. They appeared to be composed of the large amounts of slime. The increased biofilm thickness without any increase in the treatment efficiency of the system can be explained by Venkataraman and Ramanujam (1998) who reported that the substrate removal rate depends on the active outer layer of biofilm and the depth of active layer being controlled by the diffusivity of dissolved oxygen into the biofilm. The results obtained in this study were agreed with Laopaiboon et al. (2001) who found that the increased age of biofilm and consequently the increased thickness did not result in significantly higher treatment efficiency of a wastewater treatment system. The increase in the thickness was probably due to cross-linking between glutaraldehyde and the amine group of proteins or enzymes found on bacterial cell walls (Simons et al. 2000). Moreover, glutaraldehyde may also react with protein which is the one of the predominant constituents of biofilm extracellular polymeric substances (EPS) apart from polysaccharide (Christensen and Characklis, 1990; Flemming, 1993; Norwood and Gilmour, 2000).

The COD removal of the control unit (without glutaraldehyde) was 71.73 ± S.D. 7.67%. When 40 and 80 ppm glutaraldehyde were applied to the units, COD removal decreased to around 34-40 % on day 1 but then subsequently recovered and reached steady state on day 8-10 with average COD removals of 57.47 ± 1.75 and 53.05 ± 3.55% respectively. Applying at 180 ppm, glutaraldehyde caused increasing reductions in COD removal with almost total inhibition occurring on day 5 of the treatment but then slightly increased to 39.72 ± 3.10% on day 12-18. As one ppm glutaraldehyde equates 2 mg $L^{-1}$ COD, the COD of each influent wastewater containing glutaraldehyde will be increased by the presence of glutaraldehyde. When glutaraldehyde was applied to the RBCs at 40 and 80 ppm, the biocide was almost totally removed by the units since day 3 and day 8 respectively. The average concentrations of glutaraldehyde remaining in the effluents were only 5.29 ± 1.64 and 6.14 ± 1.44 ppm, respectively. At 180 ppm glutaraldehyde, the biocide content of the effluent was decreased at approximately 40 ppm between day 1 and day 5 but then subsequently decreased with glutaraldehyde concentration showing only 58.99 ± 4.11 ppm at the end of the experiment.

Glutaraldehyde at concentrations of up to 80 ppm did not significantly affect the consumption of lab-lemco broth (sole carbon source in the synthetic wastewater) by the mixed populations in the RBCs. COD removal in the untreated control RBC was around 74 mg $L^{-1}$ while at 40 and 80 ppm glutaraldehyde COD removal increased to 87 and 158 mg $L^{-1}$. The increase in COD removed at 40 and 80 ppm glutaraldehyde treatments (87 - 74 = 13 mg $L^{-1}$ and 158 - 74 = 84 mg $L^{-1}$, respectively) suggested that not only lab-lemco broth but also glutaraldehyde were removed by the RBC units. At 180 ppm glutaraldehyde treatment, the increase in COD removed (93 - 74 = 19 mg $L^{-1}$) and glutaraldehyde removed (180 - 59 = 121 ppm) indicated that the COD removed at this concentration was probably mainly derived from glutaraldehyde and not from lab-lemco broth. This implies that applying high biocide concentration to a wastewater treatment plant may inhibit simple carbon source utilization by microorganisms in the system.

*Biofilm and planktonic populations.* The colony forming measurement was chosen to assess biocide efficacy on microbial activity as it was an economical method. Glutaraldehyde had no adverse effect on the viability of the biofilms and apparent planktonic bacteria in the RBCs at concentrations of up to 80 ppm even though the thickness of the treated biofilms was greater than that of the control biofilms. The TVCs of the biofilms on the discs were similar at approximately $10^7$-$10^8$ cfu $cm^{-2}$. This implies that the biofilm thickness does not have a marked effect on the levels of efficacy of the biocide tested. Russell and Chopra (1996) reported that viability of microorganisms is not always affected when exposed to a biocide that interacts with an outer cellular component such as glutaraldehyde. Our previous study found that the maximum specific growth rate of the planktonic cells in the synthetic wastewater in a batch system was 0.84 $h^{-1}$ (Phukoetphim et al. 2004). This was far lower than the dilution rate (1.60 $h^{-1}$) in the disc chamber of the three-disc RBC operation, however, no washout of the planktonic phase in the disc chamber was observed. This indicates that the apparent planktonic population of the RBCs at all treatments including the control treatment, must be continually replenished by organisms detaching from the biofilm which serves as a continuous source of bacteria for the liquid phase (Characklis and Marshall, 1990). The TVCs of planktonic microorganisms in the RBC receiving glutaraldehyde at concentrations of up to 80 ppm were similar indicating that the number of bacteria released from the biofilms remained unchanged. Reduction in numbers of cfu's at higher glutaraldehyde concentration may be due to reduced growth rate or the death of some cells in the biofilms. Bacterial cfu's of the biofilm exposed to 180 ppm was one log reduction, implying that the growth was reduced or inhibited and consequently fewer planktonic microorganisms being detected in the disc stage. However, cfu's at 180 ppm recovered nearly to the control level following acclimation.

*Higher organisms.* Microscopic observation shows that higher organisms in the biofilms of all RBC units before the biocide treatments consisted of active protozoa and rotifers. Glutaraldehyde at concentrations of up to 80 ppm caused slight changes in population of protozoa. All higher organisms were killed in the biofilms exposed to 180 ppm glutaraldehyde since day 1 of the experiments indicating that protozoa and rotifers were more sensitive to glutaraldehyde than bacterial cells in the biofilms. The susceptibility of cells to the bactericidal activity of glutaraldehyde can be assessed by determination of MIC. MICs of glutaraldehyde on 'TSA cells' and 'control planktonic cells' were similar at 25-30 ppm. Resistance of cells to glutaraldehyde was found in 'acclimatized cells' resulting in higher MIC value. The results indicate that there was no difference in susceptibility to glutaraldehyde on cells grown in enriched medium (TSA) and the cells sloughed from biofilms which were not subjected to the biocide application. It is accepted that sessile bacteria are more resistant to a biocide than their planktonic counterparts (Eginton et al. 1998; Stewart et al. 1998; Laopaiboon et al. 2002). However, when the bacteria in the biofilms shed to the planktonic phase, they showed no increase in resistance to the biocide. After a period of acclimation, planktonic cells sloughed from the acclimatized biofilms showed significantly less susceptibility to the biocide with higher MIC when compared to those of the unacclimatized biofilms. The results also showed that the susceptibility of cells possibly related to capability of biocide degradation. Biodegradation of the acclimatized cells at 180 ppm glutaraldehyde for 2 weeks was far higher than that of the biofilms exposed to the biocide for a shorter period. The results clearly show that biocide efficacy on bacterial cells in biofilms and their planktonic counterparts are significantly different. The bacteria within biofilms are more resistant to the biocide than their planktonic cells. In planktonic cells biooxidation was totally inhibited in the presence of 25 ppm glutaraldehyde while biofilms retained some of their bio-oxidation capacity in the presence of glutaraldehyde up to 180 ppm. This may be attributed to the effect of exopolysaccharide polymers produced as a potential barrier which can prevent biocides from reaching the target organisms within the biofilms (Das et al. 1998; Morton et al. 1998; Stewart et al.1998). It can be assumed that in the presence of biocide planktonic bacteria play no part in the substrate removal and only bio-oxidation by the biofilms takes place. The results of this study indicate that there is at least a seven-fold difference in susceptibility of planktonic and biofilm cells to glutaraldehyde. The capability of the acclimatized cells to form a biofilm on the discs was investigated to simulate biofilm formation in a water system under sub-inhibitory concentrations and to study the biocide efficacy against microbial cells under these conditions. In the RBCs receiving the wastewater A and B containing 180 ppm glutaraldehyde as a sole carbon source, biofilm formation was noticed on day 3 of the experiments. This was slightly longer than in the control unit (no biocide) where the biofilm established in 2 days. The results may reflect the fact that the microorganisms were slower growing in the presence of biocide and a longer period of time was necessary for surface coverage to be achieved.

The degree of COD removal and biocide degradation depends on acclimation period, the presence of other organic matters and the amount of mineral salts available. COD and glutaraldehyde removals in the units receiving the wastewater A and B were initially quite low. After a period of acclimation for approximately 10 days, the biofilms could degrade glutaraldehyde. Glutaraldehyde removal had subsequently increased and reached steady state on day 28. The average glutaraldehyde removals were 71.36 ± 1.99 and 20.67 ± 1.94 % by the RBCs receiving the wastewater A and B respectively. Similar results were observed by COD measurement. COD removal reached constant level approximately on day 18 with average COD removal of 38.84 ± 2.93 and 19.97 ± 1.78% by the RBCs receiving the wastewater A and B respectively. The viable populations of the biofilms and the planktonic cells in the RBCs receiving the wastewater A were higher than those of the RBC receiving the wastewater B around 1.3 and 1.7 log scales respectively. However, those values were still lower when compared with those of the RBC receiving normal synthetic wastewater without glutaraldehyde. In the absence of additional mineral salts in the influent, the percentage of glutaraldehyde and COD removals was similar at approximately 20%.

This indicates that ultimate biodegradation of biocide had occurred. On the other hand, in the presence of additional mineral salts, the percentage of glutaraldehyde removal was significantly higher than that of COD removal. A possible explanation was that the biocide may not be completely degraded to $CO_2$ (mineralization) but it was converted to intermediates or combined with other components in the wastewater which could not be detected by glutaraldehyde determination but were detected as COD. However, more glutaraldehyde could have been removed if the RBC system was operated longer and/or COD loading was lower. The effect of operating conditions of the RBC performance on glutaraldehyde degradation is being investigated by the authors. The results obtained indicate that glutaraldehyde biocide could still be removed regardless the presence of simple carbon sources. However, some biocides such as isothiazolone compounds could only be degraded via co-metabolism with a growth substrate or readily degradable carbon source (Laopaiboon et al. 2001). Our previous study showed that the loss of glutaraldehyde in the RBC unit without biofilms and any reaction between the biocide and normal synthetic wastewater was minimal (Phukoetphim et al. 2004). Thus the removal of glutaraldehyde could be attributed to biological action or biodegradation. As previously mentioned, biofilms played an important role in biodegradation of the biocide because the MIC of their planktonic counterparts was much lower than the biocide concentrations applied to the RBCs. Van Ginkel (1991) and Cornelissen and Sijm (1996) reported that the difference in degree of biodegradation of certain chemicals depends on the structure of the substance, the concentration of the chemical, environmental conditions such as temperature, oxygen levels and pH and the presence of other nutrients.

In this study, pH and temperature of the wastewater in the RBC system ranged from 7.12 to 7.36 and 24.5 to 27.5°C, respectively indicating that these conditions were suitable for glutaraldehyde degradation. However, dissolved oxygen concentration in the RBC system should also be monitored to show whether there was any impact of the loading increase on DO concentration or oxygen supply was sufficient to meet the increase in COD loading. Less diversity of biofilm bacteria was observed in the RBCs receiving biocide-based wastewater. Bacterial cells from the biofilms were isolated at the end of the experiments. All tested isolates were Gram-negative rods and oxidase positive. The resistant bacterium to glutaraldehyde was tentatively identified as predominantly of the species *Burkholderia cepacia*. Other resistant species found were *Aeromonas hydrophila* and *Aeromonas salmonicida*. The results were similar to Bae and Rittmann (1995) who found that the dominant species in continuous-flow reactor of mixed phenolic compounds were *Burkhol. cepacia* and *Pseudomonas testosteroni*. Nikaido (1994) also reported that *Pseudomonas* and *Burkholderia* are highly resistant to most of the commonly used antimicrobial agents. Intrinsic resistance mechanisms, such as differences in their outer membrane permeability and the efficiency of their active efflux pumps, are thought to be responsible. In metalworking fluids, however, Gram-positive bacteria, mycobacteria, were found to be more resistant to biocides than Gram-negative bacteria, pseudomonads (Russell, 1998; Selvaraju et al. 2005). The quantity of biocides applied in, for example, cooling systems is generally calculated from MIC determined under laboratory conditions (Lutey, 1995). Due to the high cost of the volumes of biocide required and their environmental impact, addition of biocide is kept to minimum. A certain MIC of a biocide is required to maintain control over the microbial community in a system. However, in practice the active concentrations of most biocides in systems drop after addition due to a number of factors. Many biocides react with constituents of the system being treated, e.g. proteins, polysaccharides of the biofilm matrix, metal surfaces or with cells, resulting in depletion of the available biocide pool. As a result the available concentration of biocide in the system being treated is lower than required most of the time and therefore, the bacterial communities are exposed to sub-inhibitory concentrations of biocides (Brözel et al. 1995; Cloete et al. 1998). Moreover, the results indicate that susceptibility of bacteria to a biocide is reduced after a period of acclimation especially when there is an accumulation of biofilm on a surface. To reduce these effects, other modes of biocide applications such as shock treatments and use of alternating biocides should to be considered.

### Concluding Remarks

The RBC units were successfully used to simulate and predict the behaviour of a full-scale effluent treatment plant in the presence of an aldehyde biocide as well as the effect of the biocide on the activity and survival of the biofilms and planktonic cells. The biofilms became acclimated to glutaraldehyde and eventually degraded it. The degree of biocide degradation and COD removal depended on acclimation period, the presence of other organic matters and the amount of mineral salts available. No glutaraldehyde concentration used in the study was sufficiently high to eliminate microorganisms in the RBC system. Bacterial cells in biofilms were less susceptible to glutaraldehyde than their planktonic counterparts. The increased resistance of bacteria to glutaraldehyde due to acclimation should be considered in biocide applications.

### Genetically Modified Organism

A genetically modified organism (GMO) is an organism whose genetic material has been altered using techniques in genetics generally known as recombinant DNA technology. Recombinant DNA technology is the ability to combine DNA molecules from different sources into the one molecule in a test tube. Thus, the abilities or the phenotype of the organism, or the proteins it produces, can be altered through the modification of its genes. The term generally does not cover organisms whose genetic makeup has been altered by conventional cross breeding or by "mutagenesis" breeding, as these methods predate the discovery of the recombinant DNA techniques. Technically speaking, however, such techniques are, by definition, genetic modification. Examples of GMOs are diverse, and include transgenic experimental animals such as mice, several fish species, transgenic plants, or various microscopic organisms altered for the purposes of genetic research or for the production of pharmaceuticals.

The term "genetically modified organism" does not necessarily imply, but does include, transgenic substitution of genes from another species, and research is actively being conducted in this field. For example, genes for fluorescent proteins can be co-expressed with complex proteins in cultured cells to facilitate study by biologists, and modified organisms are used in researching the mechanisms of cancer and other diseases.

The first GMO was created in 1973 by Stanley N. Cohen and Herbert Boyer, demonstrating the creation of a functional organism that combined and replicated genetic information from different species. In mid-1974, very soon after the first GMO was created, scientists called for and observed a voluntary moratorium on certain recombinant DNA experiments. One goal of the moratorium was to provide time for a conference that would evaluate the state of the new technology and the risks, if any, associated with it. That conference concluded that recombinant DNA research should proceed but under strict guidelines. Such guidelines were subsequently promulgated by the National Institutes of Health in the United States and by comparable bodies in other countries. These guidelines form the basis upon which GMOs are regulated to this day. The first transgenic animals were mice created by Rudolf Jaenisch in 1974. Jaenish successfully managed to insert foreign DNA into the early-stage mouse embryos; the resulting mice carried the modified gene in all their tissues. Subsequent experiments, injecting leukemia genes to early mouse embryos using a retrovirus vector, proved the genes integrated not only to the mice themselves, but also to their progeny.

Genetic modification involves genetic engineering, also known as *gene splicing*, a technique to splice together DNA fragments from more than one organism and thus preparing a "recombinant" DNA molecule in a test tube, producing a single piece of genetic material containing the original information from multiple fragments which can then be inserted into another organism. This is achieved by cutting up DNA molecules with restriction enzymes and splicing these fragments together using DNA ligase. A *transgenic* organism that contains such DNA sequences from a foreign organism integrated into its own genome, the term "transgenic" literally means *across gene.* A mouse or fish engineered to express the green fluorescence protein, for example, would be considered a transgenic organism, since the gene coding for the protein originated from a species of jellyfish. With current technology, transgenic organisms can be produced with only a very small proportion of extraneous DNA. For example, the genome of most mammals contains three billion basepairs of DNA, while it becomes relatively difficult to insert more than 10,000 to 20,000 basepairs of foreign DNA. More sophisticated techniques using yeast artificial chromosomes and bacterial artificial chromosomes allow insertions of up to 320,000 basepairs - approximately 0.01% of the total genome.

In concept, multiple rounds of transgenesis or interbreeding of transgenics could lead to organisms with a higher proportion of foreign DNA, but cost and time considerations prevent this. In order to introduce new DNA into the receiving host, *vectors* are used. Vectors range from small circular pieces of DNA such as plasmids, to various viruses that can carry and transmit genetic information. Three processes are known by which the genetic composition of bacteria can be altered. Transformation is a process by which some bacteria are naturally capable of taking up DNA to acquire new genetic traits. This phenomenon was discovered by Frederick Griffith in 1928, although the fact that it was specifically DNA molecules that carried the genetic information was not proven until

1944. Bacteria that are competent to undergo transformation are frequently used in molecular biology. The foreign DNA uptake is facilitated by the presence of certain cations, such as $Ca^{2+}$, or by the use of electric current (electroporation). Transformation does not normally integrate new DNA into the bacterial chromosome. Instead, it remains on a plasmid. In conjugation, DNA is transferred from one bacterium to another via a temporary connecting tube of protein called a *pilus* (a process analogous to but biologically distinct from mating). A *plasmid* is transferred through the pilus. Conjugation is not widely used for the artificial genetic modification of bacteria, but happens often in nature. Transduction refers to the introduction of new DNA into a bacterial cell by a bacteriophage, a virus that infects bacteria. In order to gain knowledge about a particular gene's function, researchers often use *knock out* organisms. These organisms have a specific gene that has been functionally destroyed or "knocked out." They are used extensively in disease research with model organisms. For example, when investigating the cause of cystic fibrosis, researchers identified the CFTR gene as a likely candidate for the disease, found the mouse equivalent, bred a mouse with this gene "knocked out", and noted that the knockout mouse also had cystic fibrosis.

Like bacteria and plants, animals can be genetically modified by viral infection. However, the genetic modification occurs only in those cells that become infected, and in most cases these cells are eventually eliminated by the immune system. In some cases it is possible to use the gene-transferring ability of viruses for gene therapy, i.e. to correct diseases caused by a defective gene by supplying a normal copy of the gene. Permanent genetic modification of entire animals can be accomplished in mice. The process begins by first genetically modifying a mouse embryonic stem cell. This is normally done by physically introducing into the cell a plasmid that can integrate into the genome by a process known as transfection. During transfection the DNA integrates into the animal genome via non-homologous recombination. This altered cell is implanted into a blastocyst (an early embryo), which is then implanted into the uterus of a female mouse. A pup born from this blastocyst will be a chimera containing some cells derived from the unmodified cells of the blastocyst and some derived from the modified stem cell. By selecting mice whose germ cells (sperm- or egg-producing cells) developed from the modified cell and interbreeding them, pups that contain the genetic modification in all of their cells will be born. Baylor College of Medicine currently has one of the largest transgenic mice facilities in the country.

There has also been the genetically manipulated bull Herman with 55 offspring. A human gene was built into his genetic code while in an early embryonic stage in 1990. As a result, milk from his female descendants contained the human protein lactoferrine, that can be used as medicine, but it was present at such low levels that it was not profitable to extract them. Insects can be genetically modified by injecting them with artificial transposons and a source of the enzyme transposase. The transposon, which can include new genes, is then integrated into the genome. Such insertions are unstable and can 'jump-out' in the presence of transposase. Transgenic fish are often created by microinjection. First generation is mosaic but several lines have been produced with the transgene incorporated into the germ line and transgenic fish can then be produced "the natural way" by crossing male and female gametes. Although many types of transgenic fish exists (e.g. for increased cold tolerance, antibiotic production, ornamental Glofish etc) the main focus had been on so called growth hormone transgenic fish, mainly salmonids, tilapias and carps. These fish have an over-production of growth hormone which results in increased growth rate from a few percent up to 30-40 times that of wild-types. In some species, final size is increased as well as growth rate providing an incentive for commercial breeders to farm such fish. However, ecological concerns over potential negative effects of transgenic fish in nature largely prevent the commencement of commercial production. A large and important portion of the research on transgenic fish today is therefore focusing on environmental risk-assessment of GH-transgenic fish.

Genetic modification (GM) is the subject of controversy in its own right. Some see the science itself as intolerable meddling with "natural" order, despite many known examples of natural genetic crossings occurring throughout history (see for example horizontal gene transfer). While some would like to see it banned, others push simply for required labeling of genetically modified food. Other controversies include the definition of patent and property pertaining to products of genetic engineering and the possibility of unforeseen global side effects as a result of modified organisms proliferating. The basic ethical issues involved in genetic research are discussed in the article on genetic engineering. In 2004, Mendocino County, California became the first county in the United States to ban the production of GMOs. The measure passed with a 57% majority. In 2005, a standing committee of the government of Prince Edward Island in Canada began work to assess a proposal to ban the production of GMOs in the province. This is a largely symbolic and empty gesture as PEI has already banned GMO potatoes, which account for most of its crop. In California, the Trinity and Marin counties have also imposed bans on GM crops, while ordinances to do so were unsuccessful in Butte, San Luis Obispo, Humboldt and Sonoma counties. Supervisors in the ag-rich counties of Fresno, Kern, Kings, Solano,

Sutter and Tulare have passed resolutions supporting the practice. Currently, there is little international consensus regarding the acceptability and effective role of modified "complete" organisms such as plants or animals. A great deal of the modern research that is illuminating complex biochemical processes and disease mechanisms makes vast use of genetic engineering.

The practice of genetic modification as a scientific technique is not restricted in the United States. Individual genetically modified crops (such as soybeans) are subject to intense study before being brought to market and are common in the United States, but estimates of their market saturation vary widely. Some countries in Europe have taken the opposite position, stating that genetic modification has not been proven safe, and therefore that they will not accept genetically modified food from the United States or any other country. This issue has been brought before the World Trade Organization, which determined that not allowing modified food into the country creates an unnecessary obstacle to international trade. Consequently, genetic modification within agriculture is an issue of some strong debate in the United States, the European Union, and some other countries. Some critics have raised the concern that conventionally bred crop plants can be cross-pollinated (bred) from the pollen of modified plants. Pollen can be dispersed over large areas by wind, animals, and insects. Recent research with creeping bentgrass has lent support to the concern when modified genes were found in normal grass up to 21 km (13 miles) away from the source, and also within close relatives of the same genus Agrostis. GM proponents point out that outcrossing, as this process is known as, is not new. The same thing happens with any new open-pollinated crop variety—newly introduced traits can potentially cross out into neighbouring crop plants of the same species and, in some cases, to closely related wild relatives. Defenders of GM technology point out that each GM crop is assessed on a case by case basis to determine if there is any risk associated with the outcrossing of the GM trait into wild plant populations. The fact that a GM plant may outcross with a related wild relative is not, in itself, a risk unless such an occurrence has consequences. If, for example, a herbicide resistance trait was to cross into a wild relative of a crop plant it can be predicted that this would not have any concequences except in areas where herbicides are sprayed, such as a farm.

In such a setting the farmer can manage this risk by rotating herbicides. If patented genes are outcrossed, even accidentally, to other commercial fields and a person deliberately selects the outcrossed plants for subsequent planting then the patent holder has the right to control the use of those crops. This was supported in Canadian law in the case of Monsanto Canada Inc. v. Schmeiser.

An often cited controversy is a hypothetical Technology Protection technology (dubbed terminator by NGOs). This yet to be commercialised technology would allow the production of first generation crops that would not generate seeds in the second generation because the plants yield sterile seeds. The patent for this so-called "terminator" gene technology is owned by Delta and Pine Land and the USDA despite it often being mis-associated with Monsanto. In addition to the commercial protection of proprietary technology in selfpollinating crops such as soybean (a generally contentious issue) another purpose of the terminator gene is to prevent the escape of genetically modified traits from crosspollinating crops into wild-type species by sterilizing any resultant hybrids. The terminator gene technology created a backlash amongst those who felt the technology would prevent re-use of seed by farmers growing such terminator varieties in the developing world and was ostensibly a means to exercise patent claims. Use of the terminator technology would also prevent "volunteers", or crops that grow from unharvested seed, a major concern that arose during the Starlink debacle.

Genetically modified characters, whether as heroes, villains, or backdrop, feature prominently in many works of fiction, in particular science fiction and cyberpunk, where it is used as a plot device to explain differences in a character or setting, such as explaining increased longevity or eradication of disease in a fictional civilization. In the Spiderman movie, Peter Parker was bitten by a super-spider, enhanced with the genes of many different spiders. The abilities of all these spiders were then transferred from the super-spider, into Peter, turning him into Spiderman. The videogame character Shadow the Hedgehog was originally a science experiment who was fused with the DNA of Black Doom, causing him to have the genes of aliens as well as hedgehogs. This, however, was not revealed until the game *Shadow the Hedgehog*.

## Single Nucleotide Polymorphism (SNPs): Variations on a Theme

Wouldn't it be wonderful if you knew exactly what measures you could take to stave off, or even prevent, the onset of disease? Wouldn't it be a relief to know that you are not allergic to the drugs your doctor just prescribed? Wouldn't it be a comfort to know that the treatment regimen you are undergoing has a good chance of success because it was designed just for you? With the availability of millions of SNPs, biomedical researchers now believe that such exciting medical advances are not that far away. A SNP (pronounced "snip"), is a small genetic change, or variation, that can occur within a person's DNA sequence. The genetic code is specified by the four nucleotide "letters" A (adenine), C (cytosine), T (thymine), and G (guanine). SNP variation occurs when a single nucleotide, such as an A, replaces one of the other three nucleotide letters—C, G, or T.

An example of a SNP is the alteration of the DNA segment AAGGTTA to ATGGTTA, where the second "A" in the first snippet is replaced with a "T". On average, SNPs occur in the human population more than 1 percent of the time. Because only about 3 to 5 percent of a person's DNA sequence codes for the production of proteins, most SNPs are found outside of "coding sequences". SNPs found within a coding sequence are of particular interest to researchers because they are more likely to alter the biological function of a protein. Because of the recent advances in technology, coupled with the unique ability of these genetic variations to facilitate gene identification, there has been a recent flurry of SNP discovery and detection.

Finding single nucleotide changes in the human genome seems like a daunting prospect, but over the last 20 years, biomedical researchers have developed a number of techniques that make it possible to do just that. Each technique uses a different method to compare selected regions of a DNA sequence obtained from multiple individuals who share a common trait. In each test, the result shows a physical difference in the DNA samples only when a SNP is detected in one individual and not in the other. Many common diseases in humans are not caused by a genetic variation within a single gene but are influenced by complex interactions among multiple genes as well as environmental and lifestyle factors. Although both environmental and lifestyle factors add tremendously to the uncertainty of developing a disease, it is currently difficult to measure and evaluate their overall effect on a disease process. Therefore, we refer here mainly to a person's genetic predisposition, or the potential of an individual to develop a disease based on genes and hereditary factors. Genetic factors may also confer susceptibility or resistance to a disease and determine the severity or progression of disease. Because we do not yet know all of the factors involved in these intricate pathways, researchers have found it difficult to develop screening tests for most diseases and disorders.

By studying stretches of DNA that have been found to harbor a SNP associated with a disease trait, researchers may begin to reveal relevant genes associated with a disease. Defining and understanding the role of genetic factors in disease will also allow researchers to better evaluate the role non-genetic factors—such as behavior, diet, lifestyle, and physical activity—have on disease. Because genetic factors also affect a person's response to drug therapy, DNA polymorphisms such as SNPs will be useful in helping researchers determine and understand why individuals differ in their abilities to absorb or clear certain drugs, as well as to determine why an individual may experience an adverse side effect to a particular drug.

Therefore, the recent discovery of SNPs promises to revolutionize not only the process of disease detection but the practice of preventative and curative medicine.

### *SNPs and Disease Diagnosis*

Each person's genetic material contains a unique SNP pattern that is made up of many different genetic variations. Researchers have found that most SNPs are not responsible for a disease state. Instead, they serve as biological markers for pinpointing a disease on the human genome map, because they are usually located near a gene found to be associated with a certain disease. Occasionally, a SNP may actually cause a disease and, therefore, can be used to search for and isolate the disease-causing gene. To create a genetic test that will screen for a disease in which the disease-causing gene has already been identified, scientists collect blood samples from a group of individuals affected by the disease and analyze their DNA for SNP patterns. Next, researchers compare these patterns to patterns obtained by analyzing the DNA from a group of individuals unaffected by the disease. This type of comparison, called an "association study", can detect differences between the SNP patterns of the two groups, thereby indicating which pattern is most likely associated with the disease-causing gene. Eventually, SNP profiles that are characteristic of a variety of diseases will be established. Then, it will only be a matter of time before physicians can screen individuals for susceptibility to a disease just by analyzing their DNA samples for specific SNP patterns.

***SNPs and Drug Development***

As mentioned earlier, SNPs may also be associated with the absorbance and clearance of therapeutic agents. Currently, there is no simple way to determine how a patient will respond to a particular medication. A treatment proven effective in one patient may be ineffective in others. Worse yet, some patients may experience an adverse immunologic reaction to a particular drug. Today, pharmaceutical companies are limited to developing agents to which the "average" patient will respond. As a result, many drugs that might benefit a small number of patients never make it to market. In the future, the most appropriate drug for an individual could be determined in advance of treatment by analyzing a patient's SNP profile. The ability to target a drug to those individuals most likely to benefit, referred to as "personalized medicine", would allow pharmaceutical companies to bring many more drugs to market and allow doctors to prescribe individualized therapies specific to a patient's needs.

### SNPs and NCBI

Because SNPs occur frequently throughout the genome and tend to be relatively stable genetically, they serve as excellent biological markers. Biological markers are segments of DNA with an identifiable physical location that can be easily tracked and used for constructing a chromosome map that shows the positions of known genes, or other markers, relative to each other. These maps allow researchers to study and pinpoint traits resulting from the interaction of more than one gene. NCBI plays a major role in facilitating the identification and cataloging of SNPs through its creation and maintenance of the public SNP database (dbSNP). This powerful genetic tool may be accessed by the biomedical community worldwide and is intended to stimulate many areas of biological research, including the identification of the genetic components of disease.

### NCBI's "Discovery Space" Facilitating SNP Research

Records in dbSNP are cross-annotated within other internal information resources such as PubMed, genome project sequences, GenBank records, the Entrez Gene database, and the dbSTS database of sequence tagged sites. Users may query dbSNP directly or start a search in any part of the NCBI discovery space to construct a set of dbSNP records that satisfy their search conditions. Records are also integrated with external information resources through hypertext URLs that dbSNP users can follow to explore the detailed information that is beyond the scope of dbSNP curation. Reproduced with permission from Sherry ST, Ward MH, Kholodov M, Baker J, Phan L, Smigielski EM, Sirotkin K."dbSNP: the NCBI database of genetic variation." *Nucleic Acids Research*. 2001; 29:308-311. To facilitate research efforts, NCBI's dbSNP is included in the Entrez retrieval system which provides integrated access to a number of software tools and databases that can aid in

SNP analysis. For example, each SNP record in the database links to additional resources within NCBI's "Discovery Space".

Resources include: GenBank, NIH's sequence database; Entrez Gene, a focal point for genes and associated information; dbSTS, NCBI's resource containing sequence and mapping data on short genomic landmarks; human genome sequencing data; and PubMed, NCBI's literature search and retrieval system. SNP records also link to various external allied resources. Providing public access to a site for "one-stop SNP shopping" facilitates scientific research in a variety of fields, ranging from population genetics and evolutionary biology to large-scale disease and drug association studies. The long-term investment in such novel and exciting research promises not only to advance human biology but to revolutionize the practice of modern medicine.

#### Microarraways: An Introduction

The proper and harmonious expression of a large number of genes is a critical component of normal growth and development and the maintenance of proper health. Disruptions or changes in gene expression are responsible for many diseases.

With only a few exceptions, every cell of the body contains a full set of chromosomes and identical genes. Only a fraction of these genes are turned on, however, and it is the subset that is "expressed" that confers unique properties to each cell type. "Gene expression" is the term used to describe the transcription of the information contained within the DNA, the repository of genetic information, into messenger RNA (mRNA) molecules that are then translated into the proteins that perform most of the critical functions of cells. Scientists study the kinds and amounts of mRNA produced by a cell to learn which genes are expressed, which in turn provides insights into how the cell responds to its changing needs.

Gene expression is a highly complex and tightly regulated process that allows a cell to respond dynamically both to environmental stimuli and to its own changing needs. This mechanism acts as both an "on/off" switch to control which genes are expressed in a cell as well as a "volume control" that increases or decreases the level of expression of particular genes as necessary.

### *Enabling Technology*

Biomedical research evolves and advances not only through the compilation of knowledge but also through the development of new technologies. Using traditional methods to assay gene expression, researchers were able to survey a relatively small number of genes at a time. The emergence of new tools enables researchers to address previously intractable problems and to uncover novel potential targets for therapies. Microarrays allow scientists to analyze expression of many genes in a single experiment quickly and efficiently. They represent a major methodological advance and illustrate how the advent of new technologies provides powerful tools for researchers. Scientists are using microarray technology to try to understand fundamental aspects of growth and development as well as to explore the underlying genetic causes of many human diseases.

### DNA Microarrays

Two recent complementary advances, one in knowledge and one in technology, are greatly facilitating the study of gene expression and the discovery of the roles played by specific genes in the development of disease. As a result of the Human Genome Project, there has been an explosion in the amount of information available about the DNA sequence of the human genome. Consequently, researchers have identified a large number of novel genes within these previously unknown sequences. The challenge currently facing scientists is to find a way to organize and catalog this vast amount of information into a usable form. Only after the functions of the new genes are discovered will the full impact of the Human Genome Project be realized. The second advance may facilitate the identification and classification of this DNA sequence information and the assignment of functions to these new genes: the emergence of DNA microarray technology. A microarray works by exploiting the ability of a given mRNA molecule to bind specifically to, or hybridize to, the DNA template from which it originated. By using an array containing many DNA samples, scientists can determine, in a single experiment, the expression levels of hundreds or thousands of genes within a cell by measuring the amount of mRNA bound to each site on the array. With the aid of a computer, the amount of mRNA bound to the spots on the microarray is precisely measured, generating a profile of gene expression in the cell.

A microarray is a tool for analyzing gene expression that consists of a small membrane or glass slide containing samples of many genes arranged in a regular pattern.

Microarrays are a significant advance both because they may contain a very large number of genes and because of their small size. Microarrays are therefore useful when one wants to survey a large number of genes quickly or when the sample to be studied is small. Microarrays may be used to assay gene expression within a single sample or to compare gene expression in two different cell types or tissue samples, such as in healthy and diseased tissue. Because a microarray can be used to examine the expression of hundreds or thousands of genes at once, it promises to revolutionize the way scientists examine gene expression. This technology is still considered to be in its infancy; therefore, many initial studies using microarrays have represented simple surveys of gene expression profiles in a variety of cell types. Nevertheless, these studies represent an important and necessary first step in our understanding and cataloging of the human genome. As more information accumulates, scientists will be able to use microarrays to ask increasingly complex questions and perform more intricate experiments. With new advances, researchers will be able to infer probable functions of new genes based on similarities in expression patterns with those of known genes. Ultimately, these studies promise to expand the size of existing gene families, reveal new patterns of coordinated gene expression across gene families, and uncover entirely new categories of genes. Furthermore, because the product of any one gene usually interacts with those of many others, our understanding of how these genes coordinate will become clearer through such analyses, and precise knowledge of these inter-relationships will emerge. The use of microarrays may also speed the identification of genes involved in the development of various diseases by enabling scientists to examine a much larger number of genes. This technology will also aid the examination of the integration of gene expression and function at the cellular level, revealing how multiple gene products work together to produce physical and chemical responses to both static and changing cellular needs.

DNA Microarrays are small, solid supports onto which the sequences from thousands of different genes are immobilized, or attached, at fixed locations. The supports themselves are usually glass microscope slides, the size of two side-by-side pinky fingers, but can also be silicon chips or nylon membranes. The DNA is printed, spotted, or actually synthesized directly onto the support. The American Heritage Dictionary defines "array" as "to place in an orderly arrangement". It is important that the gene sequences in a microarray are attached to their support in an orderly or fixed way, because a researcher uses the location of each spot in the array to identify a particular gene sequence. The spots themselves can be DNA, cDNA, or oligonucleotides.

An oligonucleotide, or oligo as it is commonly called, is a short fragment of a single-stranded DNA that is typically 5 to 50 nucleotides long.

**Microarray Experiment**

One might ask, how does a scientist extract information about a disease condition from a dime-sized glass or silicon chip containing thousands of individual gene sequences? The whole process is based on hybridization probing, a technique that uses fluorescently labeled nucleic acid molecules as “mobile probes” to identify complementary molecules, sequences that are able to base-pair with one another. Each single-stranded DNA fragment is made up of four different nucleotides, adenine (A), thymine (T), guanine (G), and cytosine (C), that are linked end to end. Adenine is the complement of, or will always pair with, thymine, and guanine is the complement of cytosine. Therefore, the complementary sequence to G-T-C-C-T-A will be C-A-G-G-A-T. When two complementary sequences find each other, such as the immobilized target DNA and the mobile probe DNA, cDNA, or mRNA, they will lock together, or hybridize. Now, consider two cells: cell type 1, a healthy cell, and cell type 2, a diseased cell. Both contain an identical set of four genes, A, B, C, and D. Scientists are interested in determining the expression profile of these four genes in the two cell types. To do this, scientists isolate mRNA from each cell type and use this mRNA as templates to generate cDNA with a “fluorescent tag” attached. Different tags (red and green) are used so that the samples can be differentiated in subsequent steps. The two labeled samples are then mixed and incubated with a microarray containing the immobilized genes A, B, C, and D. The labeled molecules bind to the sites on the array corresponding to the genes expressed in each cell.

***A DNA Microarray Experiment***

1. Prepare your DNA chip using your chosen target DNAs.
2. Generate a hybridization solution containing a mixture of fluorescently labeled cDNAs.
3. Incubate your hybridization mixture containing fluorescently labeled cDNAs with your DNA chip.
4. Detect bound cDNA using laser technology and store data in a computer.
5. Analyze data using computational methods.

After this hybridization step is complete, a researcher will place the microarray in a "reader" or "scanner" that consists of some lasers, a special microscope, and a camera. The fluorescent tags are excited by the laser, and the microscope and camera work together to create a digital image of the array. These data are then stored in a computer, and a special program is used either to calculate the red-to-green fluorescence ratio or to subtract out background data for each microarray spot by analyzing the digital image of the array. If calculating ratios, the program then creates a table that contains the ratios of the intensity of red-to-green fluorescence for every spot on the array. For example, using the scenario outlined above, the computer may conclude that both cell types express gene A at the same level, that cell 1 expresses more of gene B, that cell 2 expresses more of gene C, and that neither cell expresses gene D. But remember, this is a simple example used to demonstrate key points in experimental design. Some microarray experiments can contain up to 30,000 target spots. Therefore, the data generated from a single array can mount up quickly.

### *The Colors of a Microarray*

In this schematic:

GREEN represents Control DNA, where either DNA or cDNA derived from normal tissue is hybridized to the target DNA.

RED represents Sample DNA, where either DNA or cDNA is derived from diseased tissue hybridized to the target DNA.

YELLOW represents a combination of Control and Sample DNA, where both hybridized equally to the target DNA.

BLACK represents areas where neither the Control nor Sample DNA hybridized to the target DNA.

Each spot on an array is associated with a particular gene. Each color in an array represents either healthy (control) or diseased (sample) tissue. Depending on the type of array used, the location and intensity of a color will tell us whether the gene, or mutation, is present in either the control and/or sample DNA. It will also provide an estimate of the expression level of the gene(s) in the sample and control DNA.

### Types

There are three basic types of samples that can be used to construct DNA microarrays, two are genomic and the other is "transcriptomic", that is, it measures mRNA levels. What makes them different from each other is the kind of immobilized DNA used to generate the array and, ultimately, the kind of information that is derived from the chip. The target DNA used will also determine the type of control and sample DNA that is used in the hybridization solution.

### *Changes in Gene Expression Levels*

Determining the level, or volume, at which a certain gene is expressed is called microarray expression analysis, and the arrays used in this kind of analysis are called "expression chips". The immobilized DNA is cDNA derived from the mRNA of known genes, and once again, at least in some experiments, the control and sample DNA hybridized to the chip is cDNA derived from the mRNA of normal and diseased tissue, respectively. If a gene is overexpressed in a certain disease state, then more sample cDNA, as compared to control cDNA, will hybridize to the spot representing that expressed gene. In turn, the spot will fluoresce red with greater intensity than it will fluoresce green. Once researchers have characterized the expression patterns of various genes involved in many diseases, cDNA derived from diseased tissue from any individual can be hybridized to determine whether the expression pattern of the gene from the individual matches the expression pattern of a known disease. If this is the case, treatment appropriate for that disease can be initiated. As researchers use expression chips to detect expression patterns— whether a particular gene(s) is being expressed more or less under certain circumstances—expression chips may also be used to examine changes in gene expression over a given period of time, such as within the cell cycle.

The cell cycle is a molecualr network that determines, in the normal cell, if the cell should pass through its life cycle. There are a variety of genes involved in regulating the stages of the cell cycle. Also built into this network are mechanisms designed to protect the body when this system fails or breaks down because of mutations within one of the "control genes", as is the case with cancerous cell growth. An expression microarray "experiment" could be designed where cell cycle data are generated in multiple arrays and referenced to time "zero". Analysis of the collected data could further elucidate details of the cell cycle and its "clock", providing much needed data on the points at which gene mutation leads to cancerous growth as well as sources of therapeutic intervention. In the same way, expression chips can be used to develop new drugs. For instance, if a certain gene is overexpressed in a particular form of cancer, researchers can use expression chips to see if a new drug will reduce overexpression and force the cancer into remission. Expression chips could also be used in disease diagnosis as well, e.g., in the identification of new genes involved in environmentally triggered diseases, such as those diseases affecting the immune, nervous, and pulmonary/respiratory systems.

### *Genomic Gains and Losses*

DNA repair genes are thought to be the body's frontline defense against mutations and, as such, play a major role in cancer. Mutations within these genes often manifest themselves as lost or broken chromosomes. It has been hypothesized that certain chromosomal gains and losses are related to cancer progression and that the patterns of these changes are relevant to clinical prognosis. Using different laboratory methods, researchers can measure gains and losses in the copy number of chromosomal regions in tumor cells. Then, using mathematical models to analyze these data, they can predict which chromosomal regions are most likely to harbor important genes for tumor initiation and disease progression. The results of such an analysis may be depicted as a hierarchical treelike branching diagram, referred to as a "tree model of tumor progression". Researchers use a technique called microarray Comparative Genomic Hybridization (CGH) to look for genomic gains and losses or for a change in the number of copies of a particular gene involved in a disease state. In microarray CGH, large pieces of genomic DNA serve as the target DNA, and each spot of target DNA in the array has a known chromosomal location. The hybridization mixture will contain fluorescently labeled genomic DNA harvested from both normal (control) and diseased (sample) tissue. Therefore, if the number of copies of a particular target gene has increased, a large amount of sample DNA will hybridize to those spots on the microarray that represent the gene involved in that disease, whereas comparatively small amounts of control DNA will hybridize to those same spots. As a result, those spots containing the disease gene will fluoresce red with greater intensity than they will fluoresce green, indicating that the number of copies of the gene involved in the disease has gone up.

### *Mutations in DNA*

When researchers use microarrays to detect mutations or polymorphisms in a gene sequence, the target, or immobilized DNA, is usually that of a single gene. In this case though, the target sequence placed on any given spot within the array will differ from that of other spots in the same microarray, sometimes by only one or a few specific nucleotides. One type of sequence commonly used in this type of analysis is called a Single Nucleotide Polymorphism, or SNP, a small genetic change or variation that can occur within a person's DNA sequence. Another difference in mutation microarray analysis, as compared to expression or CGH microarrays, is that this type of experiment only requires genomic DNA derived from a normal sample for use in the hybridization mixture. Once researchers have established that a SNP pattern is associated with a particular disease, they can use SNP microarray technology to test an individual for that disease expression pattern to determine whether he or she is susceptible to (at risk of developing) that disease. When genomic DNA from an individual is hybridized to an array loaded with various SNPs, the sample DNA will hybridize with greater frequency only to specific SNPs associated with that person. Those spots on the microarray will then fluoresce with greater intensity, demonstrating that the individual being tested may have, or is at risk for developing, that disease.

**Microarray Applications**

| *Microarray type* | *Application* |
|---|---|
| CGH | Tumor classification, risk assessment, and prognosis prediction |
| Expression analysis | Drug development, drug response, and therapy development |
| Mutation/Polymorphism analysis | Drug development, therapy development, and tracking disease progression |

### NCBI and Microarray Data Management

Why is it necessary to have a uniform system that will manage and provide a disbursement point for microarray data? Consider the amount of data that can potentially be generated using a single microarray chip. Suppose that chip contains 30,000 spots of target DNA. Researchers interpreting the data generated by that chip would need to know the biological identity of each target—what gene is where; the biological properties of the control and sample DNA; the experimental conditions and procedures used in setting up the experiment; and finally, the results. Although experiments such as these will undoubtedly push forward our current understanding of gene expression and regulation, many new challenges are presented in terms of data tracking and analysis. As we have just alluded, microarray technology is one of the most recent and important experimental breakthroughs in molecular biology. Today, proficiency in generating data is fast overcoming the capacity for storing and analyzing it. Much of this information is scattered across the Internet or is not even available to the public. As more laboratories acquire this technology, the problem will only get worse. This avalanche of data requires standardization of storage, sharing, and publishing techniques. To support the public use and dissemination of gene expression data, NCBI has launched the Gene Expression Omnibus, or GEO. GEO represents NCBI's effort to build an expression data repository and online resource for the storage and retrieval of gene expression data from any organism or artificial source. Many types of gene expression data, such as those types discussed in this primer, are accepted and archived as a public dataset.

### Developing MAML: Reading Off the Same Platform

Microarray Markup Language, developed by the "MAML" working group of MGED, the Microarray Gene Expression Database, is a first attempt to provide a standard platform for submitting and analyzing the enormous amounts of microarray expression data generated by different laboratories around the world. The goal of this group, which includes NCBI investigators, is to facilitate the adoption of standards for DNA-array experiment annotation and data representation, as well as the introduction of standard experimental controls and data normalization methods. The underlying goal is to facilitate the establishment of gene expression data repositories, the comparability of gene expression data from different sources, the interoperability of different gene expression databases, and data analysis software. MAML proposes a framework for describing information about a DNA-array experiment and a data format for communicating this information, including details about:

- *Experimental design:* the set of the hybridization experiments as a whole
- *Array design:* each array used and each spot on the array
- *Samples:* samples used, the extract preparation, and labeling

- *Hybridizations:* procedures and parameters
- *Measurements:* images, quantitation, and specifications
- *Controls:* types, values, and specifications

MAML is independent of the particular experimental platform and provides a framework for describing experiments done on all types of DNA arrays, including spotted and synthesized arrays, as well as oligo and cDNA arrays. What's more, MAML provides format to represent microarray data in a flexible way, which allows analysis of data obtained from not only any existing microarray platforms but also many of the possible future variants, including protein arrays. Although the data in GEO are not currently provided in MAML format, it is NCBI's goal to have the data delivered in a number of formats, including MAML, soon to be replaced by a more recent version called MAGEML (MicroArray Gene Expression Markup Language).

#### *The Benefit of GEO and MAML*

- By storing vast amounts of data on gene expression profiles derived from multiple experiments using varied criteria and conditions, GEO will aid in the study of functional genomics—the development and application of global experimental approaches to assess gene function
- GEO will facilitate the cross-validation of data obtained using different techniques and technologies and will help set benchmarks and standards for further gene expression studies
- By making the information stored in GEO publicly available, the fields of bioinformatics and functional genomics will be both promoted and advanced
- That such experimental data should be freely

accessible to all is consistent with NCBI's legislative mandate and mission: to develop new information technologies to aid in the understanding of fundamental molecular and genetic processes that control health and disease

**Microarray Technology: Future Roles in Treating Diseases**

Now that you understand the concept behind array technology, picture this: a hand-held instrument that a physician could use to quickly diagnose cancer or other diseases during a routine office visit. What if that same instrument could also facilitate a personalized treatment regimen, exactly right for you? Personalized drugs. Molecular diagnostics. Integration of diagnosis and therapeutics.

These are the long-term promises of microarray technology. Maybe not today or even tomorrow, but someday. For the first time, arrays offer hope for obtaining global views of biological processes—simultaneous readouts of all the body's components—by providing a systematic way to survey DNA and RNA variation. NCBI, by continuing its efforts to provide a standard format for microarray data and to provide free, universal access to that data, will help the scientific community in making those promises realities.

# 4

# Theoretical and Practical Perspectives on Biotechnology

## An Introduction to Theoretical Perspectives

Since the 1960s, legal theory has expanded to take into account insights drawn from other disciplines, especially economics and sociology. Economic approaches to law have greatly enriched the theoretical framework of legal thought, while sociological approaches have emphasised the importance of empirical data as a basis for sound legal policy. Because the strengths and weaknesses of these two styles of legal thinking tend to complement each other, combining economic and sociological perspectives on particular legal issues can greatly enhance understanding of those issues. (The rapid development in the late 1990s of "law and norms" theory, an area of legal scholarship influenced by "new institutional economics", or NIE, exemplifies attempts to incorporate both economic and sociological insights into legal analysis. Law and norms theory has been applied to questions of intellectual property protection and open communication in scientific research). An interdisciplinary approach makes particular sense in relation to questions about intellectual property protection for biotechnology research tools because both economics and sociology have traditionally had something to say about intellectual property rights. Traditional economic theories tended to support strong intellectual property rights in general, while traditional sociology of science theories emphasised the importance of free and open scientific exchange. More recent theorising in both disciplines is more nuanced, but tends to emphasise the importance of information flow for ongoing innovation and the existence of uncertainties and obstacles (transaction costs) associated with the exchange of scientific information. This section outlines traditional and more recent theories about the relationship between intellectual property, information flow and the progress of scientific research in both sociology and economics literature. It then traces the legal literature on research tools in biotechnology from its emergence in response to dramatic changes in the US intellectual property system in 1980s to its incorporation into a wider legal literature on the nature and role of the "intellectual commons" (and the "public domain") at the turn of the twentieth century. Finally, we note the emergence of another substrand of that wider legal literature that analyses the Free and Open Source software revolution of the 1980s and 1990s and attempts to generalise the analysis to a concept of "commons-based peer production".

### Gene Discovery, Expression and ESTs

Investigators are working diligently to sequence and assemble the genomes of various organisms, including the mouse and human, for a number of important reasons. Although important goals of any sequencing project may be to obtain a genomic sequence and identify a complete set of genes, the ultimate goal is to gain an understanding of when, where, and how a gene is turned on, a process commonly referred to as gene expression. Once we begin to understand where and how a gene is expressed under normal circumstances, we can then study what happens in an altered state, such as in disease. To accomplish the latter goal, however, researchers must identify and study the protein, or proteins, coded for by a gene. As one can imagine, finding a gene that codes for a protein, or proteins, is not easy. Traditionally, scientists would start their search by defining a biological problem and developing a strategy for researching the problem. Oftentimes, a search of the scientific literature provided various clues about how to proceed. For example, other laboratories may have published data that established a link between a particular protein and a disease of interest. Researchers would then work to isolate that protein, determine its function, and locate the gene that coded for the protein. Alternatively, scientists could conduct what is referred to as linkage studies to determine the chromosomal location of a particular gene. Once the chromosomal location was determined, scientists would use biochemical methods to isolate the gene and its corresponding protein. Either way, these methods took a great deal of time—years in some cases—and yielded the location and description of only a small percentage of the genes found in the human genome. Now, however, the time required to locate and fully describe a gene is rapidly decreasing, thanks to the development of, and access to, a technology used to generate what are called Expressed Sequence Tags, or ESTs. ESTs provide researchers with a quick and inexpensive route for discovering new genes, for obtaining data on gene expression and regulation, and for constructing genome maps. Today, researchers using ESTs to study the human genome find themselves riding the crest of a wave of scientific discovery the likes of which has never been seen before.

> An Expressed Sequence Tag is a tiny portion of an entire gene that can be used to help identify unknown genes and to map their positions within a genome.

ESTs are small pieces of DNA sequence (usually 200 to 500 nucleotides long) that are generated by sequencing either one or both ends of an expressed gene. The idea is to sequence bits of DNA that represent genes expressed in certain cells, tissues, or organs from different organisms and use these "tags" to fish a gene out of a portion of chromosomal DNA by matching base pairs. The challenge associated with identifying genes from genomic sequences varies among organisms and is dependent upon genome size as well as the presence or absence of introns, the intervening DNA sequences interrupting the protein coding sequence of a gene. Gene identification is very difficult in humans, because most of our genome is composed of introns interspersed with a relative few DNA coding sequences, or genes. These genes are expressed as proteins, a complex process composed of two main two steps. Each gene (DNA) must be converted, or transcribed, into messenger RNA (mRNA), RNA that serves as a template for protein synthesis. The resulting mRNA then guides the synthesis of a protein through a process called translation. Interestingly, mRNAs in a cell do not contain sequences from the regions between genes, nor from the non-coding introns that are present within many genes. Therefore, isolating mRNA is key to finding expressed genes in the vast expanse of the human genome. The problem, however, is that mRNA is very unstable outside of a cell; therefore, scientists use special enzymes to convert it to complementary DNA (cDNA). cDNA is a much more stable compound and, importantly, because it was generated from a mRNA in which the introns have been removed, cDNA represents only expressed DNA sequence.

> cDNA is a form of DNA prepared in the laboratory using an enzyme called reverse transcriptase. cDNA production is the reverse of the usual process of transcription in cells because the procedure uses mRNA as a template rather than DNA. Unlike genomic DNA, cDNA contains only expressed DNA sequences, or exons.

### From cDNAs to ESTs

A *"gene family"* is a group of closely related genes that produces similar protein products.

A UTR is that part of a gene that is not translated into protein.

Once cDNA representing an expressed gene has been isolated, scientists can then sequence a few hundred nucleotides from either end of the molecule to create two different kinds of ESTs. Sequencing only the beginning portion of the cDNA produces what is called a 5' EST. A 5' EST is obtained from the portion of a transcript that usually codes for a protein. These regions tend to be conserved across species and do not change much within a gene family. Sequencing the ending portion of the cDNA molecule produces what is called a 3' EST. Because these ESTs are generated from the 3' end of a transcript, they are likely to fall within non-coding, or untranslated regions (UTRs), and therefore tend to exhibit less cross-species conservation than do coding sequences. A UTR is the part of a gene that is not translated into protein.

### ESTs as Genome Landmarks

Just as a person driving a car may need a map to find a destination, scientists searching for genes also need genome maps to help them to navigate through the billions of nucleotides that make up the human genome. For a map to make navigational sense, it must include reliable landmarks or "markers". Currently, the most powerful mapping technique, and one that has been used to generate many genome maps, relies on Sequence Tagged Site (STS) mapping. An STS is a short DNA sequence that is easily recognizable and occurs only once in a genome (or chromosome). The 3' ESTs serve as a common source of STSs because of their likelihood of being unique to a particular species and provide the additional feature of pointing directly to an expressed gene.

### ESTs as Gene Discovery Resources

**ESTs are powerful tools in the hunt for known genes because they greatly reduce the time required to locate a gene.**

Because ESTs represent a copy of just the interesting part of a genome, that which is expressed, they have proven themselves again and again as powerful tools in the hunt for genes involved in hereditary diseases. ESTs also have a number of practical advantages in that their sequences can be generated rapidly and inexpensively, only one sequencing experiment is needed per each cDNA generated, and they do not have to be checked for sequencing errors because mistakes do not prevent identification of the gene from which the EST was derived.

**Using ESTs, scientists have rapidly isolated some of the genes involved in Alzheimer's disease and colon cancer.**

To find a disease gene using this approach, scientists first use observable biological clues to identify ESTs that may correspond to disease gene candidates. Scientists then examine the DNA of disease patients for mutations in one or more of these candidate genes to confirm gene identity. Using this method, scientists have already isolated genes involved in Alzheimer's disease, colon cancer, and many other diseases. It is easy to see why ESTs will pave the way to new horizons in genetic research.

### ESTs and NCBI

**For ESTs to be easily accessed and useful as gene discovery tools, they must be organized in a searchable database that also provides access to genome data.**

Because of their utility, speed with which they may be

generated, and the low cost associated with this technology, many individual scientists as well as large genome sequencing centers have been generating hundreds of thousands of ESTs for public use. Once an EST was generated, scientists were submitting their tags to GenBank, the NIH sequence database operated by NCBI. With the rapid submission of so many ESTs, it became difficult to identify a sequence that had already been deposited in the database. It was becoming increasingly apparent to NCBI investigators that if ESTs were to be easily accessed and useful as gene discovery tools, they needed to be organized in a searchable database that also provided access to other genome data. Therefore, in 1992, scientists at NCBI developed a new database designed to serve as a collection point for ESTs. Once an EST that was submitted to GenBank had been screened and annotated, it was then deposited in this new database, called dbEST.

### dbEST: A Descriptive Catalog of ESTs

**Scientists at NCBI annotate EST records with text information regarding DNA and mRNA homologies.**

Scientists at NCBI created dbEST to organize, store, and provide access to the great mass of public EST data that has already accumulated and that continues to grow daily. Using dbEST, a scientist can access not only data on human ESTs but information on ESTs from over 300 other organisms as well. Whenever possible, NCBI scientists annotate the EST record with any known information. For example, if an EST matches a DNA sequence that codes for a known gene with a known function, that gene's name and function are placed on the EST record. Annotating EST records allows public scientists to use dbEST as an avenue for gene discovery. By using a database search tool, such as NCBI's BLAST, any interested party can conduct sequence similarity searches against dbEST.

### UniGene

Because a gene can be expressed as mRNA many, many times, ESTs ultimately derived from this mRNA may be redundant. That is, there may be many identical, or similar, copies of the same EST. Such redundancy and overlap means that when someone searches dbEST for a particular EST, they may retrieve a long list of tags, many of which may represent the same gene. Searching through all of these identical ESTs can be very time consuming. To resolve the redundancy and overlap problem, NCBI investigators developed the UniGene database UniGene automatically partitions GenBank sequences into a non-redundant set of gene-oriented clusters. Although it is widely recognized that the generation of ESTs constitutes an efficient strategy to identify genes, it is important to acknowledge that despite its advantages, there are several limitations associated with the EST approach. One is that it is very difficult to isolate mRNA from some tissues and cell types. This results in a paucity of data on certain genes that may only be found in these tissues or cell types. Second is that important gene regulatory sequences may be found within an intron. Because ESTs are small segments of cDNA, generated from a mRNA in which the introns have been removed, much valuable information may be lost by focusing only on cDNA sequencing. Despite these limitations, ESTs continue to be invaluable in characterizing the human genome, as well as the genomes of other organisms. They have enabled the mapping of many genes to chromosomal sites and have also assisted in the discovery of many new genes.

### Molecular Modeling

Proteins are fundamental components of all living cells. They exhibit an enormous amount of chemical and structural diversity, enabling them to carry out an extraordinarily diverse range of biological functions. Proteins help us digest our food, fight infections, control body chemistry, and in general, keep our bodies functioning smoothly. Scientists know that the critical feature of a protein is its ability to adopt the right shape for carrying out a particular function. But sometimes a protein twists into the wrong shape or has a missing part, preventing it from doing its job. Many diseases, such as Alzheimer's and "mad cow", are now known to result from proteins that have adopted an incorrect structure. Identifying a protein's shape, or structure, is key to understanding its biological function and its role in health and disease. Illuminating a protein's structure also paves the way for the development of new agents and devices to treat a disease. Yet solving the structure of a protein is no easy feat. It often takes scientists working in the laboratory months, sometimes years, to experimentally determine a single structure. Therefore, scientists have begun to turn toward computers to help predict the structure of a protein based on its sequence. The challenge lies in developing methods for accurately and reliably understanding this intricate relationship.

**Levels of Protein Structure**

To produce proteins, cellular structures called ribosomes join together long chains of subunits. A set of 20 different subunits, called amino acids, can be arranged in any order to form a polypeptide that can be thousands of amino acids long. These chains can then loop about each other, or fold, in a variety of ways, but only one of these ways allows a protein to function properly. The critical feature of a protein is its ability to fold into a conformation that creates structural features, such as surface grooves, ridges, and pockets, which allow it to fulfill its role in a cell. A protein's conformation is usually described in terms of levels of structure. Traditionally, proteins are looked upon as having four distinct levels of structure, with each level of structure dependent on the one below it. In some proteins, functional diversity may be further amplified by the addition of new chemical groups after synthesis is complete.

The stringing together of the amino acid chain to form a polypeptide is referred to as the primary structure. The secondary structure is generated by the folding of the primary sequence and refers to the path that the polypeptide backbone of the protein follows in space. Certain types of secondary structures are relatively common. Two well-described secondary structures are the alpha helix and the beta sheet. In the first case, certain types of bonding between groups located on the same polypeptide chain cause the backbone to twist into a helix, most often in a form known as the alpha helix. Beta sheets are formed when a polypeptide chain bonds with another chain that is running in the opposite direction. Beta sheets may also be formed between two sections of a single polypeptide chain that is arranged such that adjacent regions are in reverse orientation. The tertiary structure describes the organization in three dimensions of all of the atoms in the polypeptide. If a protein consists of only one polypeptide chain, this level then describes the complete structure. Multimeric proteins, or proteins that consist of more than one polypeptide chain, require a higher level of organization. The quaternary structure defines the conformation assumed by a multimeric protein. In this case, the individual polypeptide chains that make up a multimeric protein are often referred to as the protein subunits. The four levels of protein structure are hierarchal, that is, each level of the build process is dependent upon the one below it.

A protein's primary amino acid sequence is crucial in determining its final structure. In some cases, amino acid sequence is the sole determinant, whereas in other cases, additional interactions may be required before a protein can attain its final conformation. For example, some proteins require the presence of a cofactor, or a second molecule that is part of the active protein, before it can attain its final conformation. Multimeric proteins often require one or more subunits to be present for another subunit to adopt the proper higher order structure. In any case, as we stated earlier, the entire process is cooperative, that is, the formation of one region of secondary structure determines the formation of the next region.

### Allosteric Proteins

Under certain conditions, a protein may have a stable alternate conformation, or shape, that enables it to carry out a different biological function. Proteins that exhibit this characteristic are called allosteric. The interaction of an allosteric protein with a specific cofactor, or with another protein, may influence the transition of the protein between shapes. In addition, any change in conformation brought about by an interaction at one site may lead to an alteration in the structure, and thus function, at another site. One should bear in mind, though, that this type of transition affects only the protein's shape, not the primary amino acid sequence. Allosteric proteins play an important role in both metabolic and genetic regulation. Traditionally, a protein's structure was determined using one of two techniques: X-ray crystallography or nuclear magnetic resonance (NMR) spectroscopy.

### X-ray Crystallography

Crystals are a solid form of a substance in which the component molecules are present in an ordered array called a lattice. The basic building block of a crystal is called a unit cell. Each unit cell contains exactly one unique set of the crystal's components, the smallest possible set that is fully representative of the crystal. Crystals of a complex molecule, like a protein, produce a complex pattern of X-ray diffraction, or scattering of X-rays. When the crystal is placed in an X-ray beam, all of the unit cells present the same face to the beam; therefore, many molecules are in the same orientation with respect to the incoming X-rays. The X-ray beam enters the crystal and a number of smaller beams emerge: each one in a different direction, each one with a different intensity. If an X-ray detector, such as a piece of film, is placed on the opposite side of the crystal from the X-ray source, each diffracted ray, called a reflection, will produce a spot on the film. However, because only a few reflections can be detected with any one orientation of the crystal, an important component of any X-ray diffraction instrument is a device for accurately setting and changing the orientation of the crystal.

The set of diffracted, emerging beams contains information about the underlying crystal structure. If we could use light instead of X-rays, we could set up a system of lenses to recombine the beams emerging from the crystal and thus bring into focus an enlarged image of the unit cell and the molecules therein. But the molecules do not diffract visible light, and X-rays, unlike light, cannot be focused with lenses. However, the scientific laws that lenses obey are well understood, and it is possible to calculate the molecular image with a computer. In effect, the computer mimics the action of a lens. The major drawback associated with this technique is that crystallization of the proteins is a difficult task. Crystals are formed by slowly precipitating proteins under conditions that maintain their native conformation or structure. These exact conditions can only be discovered by repeated trials that entail varying certain experimental conditions, one at a time. This is a very time consuming and tedious process. In some cases, the task of crystallizing a protein borders on the impossible.

### Nuclear Magnetic Resonance (NMR) Spectroscopy

The basic phenomenon of NMR spectroscopy was discovered in 1945. In this technique, a sample is immersed in a magnetic field and bombarded with radio waves. These radio waves encourage the nuclei of the molecule to resonate, or spin. As the positively charged nucleus spins, the moving charge creates what is called a magnetic moment. The thermal motion of the molecule—the movement of the molecule associated with the temperature of the material—further creates a torque, or twisting force, that makes the magnetic moment "wobble" like a child's top. When the radio waves hit the spinning nuclei, they tilt even more, sometimes flipping over. These resonating nuclei emit a unique signal that is then picked up on a special radio receiver and translated using a decoder. This decoder is called the Fourier Transform algorithm, a complex equation that translates the language of the nuclei into something a scientist can understand. By measuring the frequencies at which different nuclei flip, scientists can determine molecular structure, as well as many other interesting properties of the molecule. In the past 10 years, NMR has proven to be a powerful alternative to X-ray crystallography for the determination of molecular structure. NMR has the advantage over crystallographic techniques in that experiments are performed in solution as opposed to a crystal lattice. However, the principles that make NMR possible tend to make this technique very time consuming and limit the application to small- and medium-sized molecules.

### Computational Modeling

Researchers have been working for decades to develop procedures for predicting protein structure that are not so time consuming and that are not hindered by size and solubility constraints. To do this, researchers have turned to computers for help in predicting protein structure from gene sequences, a concept called homology modeling. The complete genomes of various organisms, including humans, have now been decoded and allow researchers to approach this goal in a logical and organized fashion. Before we go any further, it is important to define some common terminology used in this field.

### Some Basic Theory

It is theorized that proteins that share a similar sequence generally share the same basic structure. Therefore, by experimentally determining the structure for one member of a protein family, called a target, researchers have a model on which to base the structure of other proteins within that family. Moving a step further, by selecting a target from each superfamily, researchers can study the universe of protein folds in a systematic fashion and outline a set of sequences associated with each folding motif. Many of these sequences may not demonstrate a resemblance to one another, but their identification and assignment to a particular fold is essential for predicting future protein structures using homology modeling. The scientific basis for these theories is that a strong conservation of protein three-dimensional shape across large evolutionary distances—both within single species, between species, and in spite of sequence variation—has been demonstrated again and again. Although most scientists choose high-priority structures as their targets, this theory provides the option to choose any one of the proteins within a family as the target, rather than trying to achieve experimental results using a protein that is particularly difficult to work with using crystallographic or NMR techniques.

### Maximizing Results

Specific tasks must be carried out to maximize results when determining protein structure using homology modeling. First, protein sequences must be organized in terms of families, preferably in a searchable database, and a target must be selected. Protein families can be identified and organized by comparing protein sequences derived from completely sequenced genomes. Targets may be selected for families that do not exhibit apparent sequence homology to proteins with a known three-dimensional structure. Next, researchers must generate a purified protein for analysis of the chosen target and then experimentally determine the target's structure, either by X-ray crystallography and/or NMR. Target structures determined experimentally may then be further analyzed to evaluate their similarity to other known protein structures and to determine possible evolutionary relationships that are not identifiable from protein sequence alone. The target structure will also serve as a detailed model for determining the structure of other proteins within that family. In favorable cases, just knowing the structure of a particular protein may also provide considerable insight into its possible function.

### The Molecular Modeling Database

NCBI's Molecular Modeling DataBase (MMDB), an integral part of our Entrez information retrieval system, is a compilation of all of the PDB three-dimensional structures of biomolecules. The difference between the two databases is that the MMDB records reorganize and validate the information stored in the database in a way that enables cross-referencing between the chemistry and the three-dimensional structure of macromolecules. By integrating chemical, sequence, and structure information, MMDB is designed to serve as a resource for structure-based homology modeling and protein structure prediction

NCBI has also developed a three-dimensional structure viewer, called Cn3D, for easy interactive visualization of molecular structures from Entrez. Cn3D serves as a visualization tool for sequences and sequence alignments. What sets Cn3D apart from other software is its ability to correlate structure and sequence information. For example, using Cn3D, a scientist can quickly locate the residues in a crystal structure that correspond to known disease mutations or conserved active site residues from a family of sequence homologs, or sequences that share a common ancestor. Cn3D displays structure-structure alignments along with the corresponding structure-based sequence alignments to emphasize those regions within a group of related proteins that are most conserved in structure and sequence. Cn3D also features custom labeling options, high-quality graphics, and a variety of file exports that together make Cn3D a powerful tool for literature annotation.

### Protein Data Bank

The PDB was the first "bioinformatics" database ever built and is designed to store complex three-dimensional data. The PDB was originally developed and housed at the Brookhaven National Laboratories but is now managed and maintained by the Research Collaboratory for Structural Bioinformatics (RCSB). The PDB is a collection of all publicly available three-dimensional structures of proteins, nucleic acids, carbohydrates, and a variety of other complexes experimentally determined by X-ray crystallography and NMR.

### PDBeast: Taxonomy in MMDB

Taxonomy is the scientific discipline that seeks to catalog and reconstruct the evolutionary history of life on earth. NCBI's Structure Group, in collaboration with NCBI's taxonomists, has undertaken taxonomy annotation for the structure data stored in MMDB. A semi-automated approach has been implemented in which a human expert checks, corrects, and validates automatic taxonomic assignments. The PDBeast software tool was developed by NCBI for this purpose. It pulls text-descriptions of "Source Organisms" from either the original PDB-Entries or user-specified information and looks for matches in the NCBI Taxonomy database to record taxonomy assignments.

### Structure Similarity Searching Using VAST

As just noted, a sequence-sequence similarity program provides an alignment of two sets of sequences. A structure-structure similarity program provides a three-dimensional structure superposition. Structure similarity search services are based on the premise that some measure can be computed between two structures to assess their similarities, much the same way a BLAST alignment is predicted. VAST or the Vector Alignment Search Tool, is a computer algorithm developed at NCBI for use in identifying similar three-dimensional protein structures. VAST is capable of detecting structural similarities between proteins stored in MMDB, even when no sequence similarity is detected. VAST Search is NCBI's structure-structure similarity search service that compares three-dimensional coordinates of newly determined protein structures to those in the MMDB or PDB databases. VAST Search creates a list of structure neighbors, or related structures, that a user can then browse interactively. VAST Search will retrieve almost all structures with an identical three-dimensional fold, although it may occasionally miss a few structures or report chance similarities.

### COGs: Phylogenetic Classification of Proteins

The database of Clusters of Orthologous Groups of proteins (COGs) represents an attempt at the phylogenetic classification of proteins— a scheme that indicates the evolutionary relationships between organisms—from complete genomes. Each COG includes proteins that are thought to be orthologous. Orthologs are genes in different species derived from a common ancestor and carried on through evolution. COGs may be used to detect similarities and differences between species for identifying protein families and predicting new protein functions and to point to potential drug targets in disease-causing species. The database is accompanied by the COGnitor program, which assigns new proteins, typically from newly sequenced genomes, to pre-existing COGs. A Web page containing additional structural and functional information is now associated with each COG. These hyperlinked information pages include: systematic classification of the COG members under the different classification systems; indications of which COG members (if any) have been characterized genetically and biochemically; information on the domain architecture of the proteins constituting the COG and the three-dimensional structure of the domains if known or predictable; a succinct summary of the common structural and functional features of the COG members as well as peculiarities of individual members; and key references.

### Detecting New Sequence Similarities: BLAST against MMDB

Comparison, whether of structural features or protein sequences, lies at the heart of biology. The introduction of BLAST, or The Basic Local Alignment Search Tool, in 1990 made it easier to rapidly scan huge databases for overt homologies, or sequence similarities, and to statistically evaluate the resulting matches. BLAST works by comparing a user's unknown sequence against the database of all known sequences to determine likely matches. Sequence similarities found by BLAST have been critical in several gene discoveries. Hundreds of major sequencing centers and research institutions around the country use this software to transmit a query sequence from their local computer to a

BLAST server at the NCBI via the Internet. In a matter of seconds, the BLAST server compares the user's sequence with up to a million known sequences and determines the closest matches. Not all significant homologies are readily and easily detected, however. Some of the most interesting are subtle similarities that do not always rise to statistical significance during a standard BLAST search. Therefore, NCBI has extended the statistical methodology used in the original BLAST to address the problem of detecting weak, yet significant, sequence similarities. PSI-BLAST, or Position-Specific Iterated BLAST, searches sequence databases with a profile constructed using BLAST alignments, from which it then constructs what is called a position-specific score matrix. For protein analysis, the new Pattern Hit Initiated BLAST, or PHI-BLAST, serves to complement the profile-based searching that was previously introduced with PSI-BLAST. PHI-BLAST further incorporates hypotheses as to the biological function of a query sequence and restricts the analysis to a set of protein sequences that is already known to contain a specific pattern or motif.

### Application to Biomedicine

Although the information derived from modeling studies is primarily about molecular function, protein structure data also provide a wealth of information on mechanisms linked to the function and the evolutionary history of and relationships between macromolecules. NCBI's goals in adding structure data to its Web site are to make this information easily accessible to the biomedical community worldwide and to facilitate comparative analysis involving three-dimensional structure.

### The Conserved Domain Database

The Conserved Domain Database (CDD) is a collection of sequence alignments and profiles representing protein domains conserved in molecular evolution. It includes domains from SMART and Pfam, two popular Web-based tools for studying sequence domains, as well as domains contributed by NCBI researchers. CD Search, another NCBI search service, can be used to identify conserved domains in a protein query sequence. CD-Search uses RPS-BLAST to compare a query sequence against specific matrices that have been prepared from conserved domain alignments present in CDD. Alignments are also mapped to known three-dimensional structures and can be displayed using Cn3D (see above).

### Conserved Domain Architecture Retrieval Tool

NCBI's Conserved Domain Architecture Retrieval Tool (CDART) displays the functional domains that make up a protein and lists other proteins with similar domain architectures. CDART determines the domain architecture of a query protein sequence by comparing it to the CDD, a database of conserved domain alignments, using RPS-BLAST. The Conserved Domain Architecture Retrieval Tool then compares the protein's domain architecture to that of other proteins in NCBI's non-redundant sequence database. Related sequences are identified as those proteins that share one or more similar domains. CDART displays these sequences using a graphical summary that depicts the types and locations of domains identified within each sequence. Links to the individual sequences as well as to further information on their domain architectures are also provided. Because protein domains may be considered elementary units of molecular function and proteins related by domain architecture may play similar roles in cellular processes, CDART serves as a useful tool in comparative sequence analysis.

# Bibliography

Abraham, E. J., et al., "Insulinotropic hormone glucagons-like peptide-1 differentiation of human pancreatic islet-derived progenitor cells into insulinproducing cells," Endocrinology 143: 3152-3161 (2002).

Alan Meisel of the University of Pittsburgh School of Law, quoted in Coyle, M., "The Clone Zone," *The National Law Journal*, May 15, 2002.

Asahara T, Takahashi T, Masuda H, Kalka C, Chen D, Iwaguro H, Inai Y, Silver M, Isner JM. VEGF contributes to postnatal neovascularization by mobilizing bone marrow-derived endothelial progenitor cells. EMBO J 1999; 18:3964- 3972.

Asahara, T. et al. Isolation of putative progenitor cells for endothelial angiogenesis. Science 1997; 275:964-967.

Assady, S., *et al.*, "Insulin production by human embryonic stem cells," Diabetes 50: 1691-1697 (2001).

Atkinson, M. A. and Leiter, E. H., "The NOD mouse model of type 1 diabetes: As good as it gets?" Nature Medicine 5: 601-604 (1999).

Baker, K. S., *et al.*, "New Malignancies After Blood or Marrow Stem-Cell Transplantation in Children and Adults: Incidence and Risk Factors," Journal of Clinical Investigation 21: 1352-1358 (2003).

Brivanlou, A. H., *et al.*, "Stem cells. Setting standards for human embryonic stem cells," Science 300: 913-916 (2003).

Bush, G.W., "Stem Cell Science and the Preservation of Life," *The New York Times*, August 12, 2001, p. D13.

Callahan, D., *What Price Better Health: The Hazards of the Research Imperative.*

Carpenter, Janet and Leonard Gianessi, "Herbicide Tolerant Soybeans: Why Growers Are Adopting Roundup Ready Varieties," *AgBioForum*, vol. 2, no. 2, 1999. Available at www.agbioforum.missouri.edu.

Carpenter, M.K., et al., "Characterization and Differentiation of Human Embryonic Stem Cells," Cloning and Stem Cells 5: 79-88 (2003), and Pittenger, *M. F. et al., "Multil*ineage potential of adult human mesenchymal stem cells," Science 284: 143-147 (1999).

Chen, Y., *et al.*, "Embryonic stem cells generated by nuclear transfer of human somatic nuclei into rabbit oocytes," Cell Research 13: 251-264 (2003).

Childress, J., "Federal Policy Toward Embryonic Stem cell research," *American Journal of Bioethics* 2(1): 34 (2002).

Choi K, Kennedy M, Kazarov A, Papadimitriou, Keller G. A common precursor for hematopoietic and endothelial cells. Development 1998;125:725-732.

Chu, K., et *al.*, "Human neural stem cells can migrate, differentiate, and integrate after intravenous transplantation in adult rats with transient forebrain ischemia," Neuroscience Letters 343: 129-133 (2003).

Clark, J., "Squandering Our Technological Future," *The New York Times*, August 30, 2001, p. A19. See also the presentation of Thomas Okarma, President and CEO of Geron Corporation, before the Council on September 4, 2003.

Conference of Catholic Bishops, in a news release ("Catholic Bishops Criticize Bush Policy on Embryo Research," made available by the United States Conference of Catholic Bishops, August 9, 2001).

Daley, G. Q., *et al.*, "Realistic Prospects for Stem Cell Therapeutics," Hematology American Society for Hematology Education *Program: 39*8-418 (2003).

Daley, G., "Cloning and Stem Cells—Handicapping the Political and Scientific Debates," *New England Journal of Medicine* 349(3): 211-212 (2003).

Doerflinger, R., "Ditching Religion and Reality," *American Journal of Bioethics* 2(1): 31 (2002).

Drukker, M., et al., "Characterization of the expression of MHC proteins in human embryonic stem cells," Proceedings of the National Academy of Sciences of the *United States of America 99: 9864-9869 (2002).*

Elefanty AG, Robb L, Birner R, Begley CG. Hematopoietic-specific genes are not induced during in vitro differentiation of scl-null embryonic stem cells.

Englund, U., *et al.,* "Transplantation of human neural progenitor cells into the neonatal rat brain: extensive migration and differentiation with longdistance axonal projections," Experimental Neurology 173: 1-21 (2002).

Faden, *et al.,* "Public Stem Cell Banks: Considerations of Justice in Stem Cell Research and Therapy" *Hastings Center Report* 33:6 (2003).

Faraci, M., *et al.,* "Severe neurologic complications after hematopoietic stem cell transplantation in children," Neurology 59: 1895-1904 (2002).

Folkman, J. Therapeutic angiogenesis in ischemic limbs. Circulation 1998; 97:108- 110.

Gianessi, Leonard P. and Janet E. Carpenter, "Agricultural Biotechnology: Insect Control Benefits," National Center for Food and Agricultural Policy, 1999. Available on Biotechnology Industry Organization at www.bio.org.

Hogan, P., et al., "Economic Costs of Diabetes in the US in 2002," Diabetes Care 26: 917-932 (2003).

Hori, Y., et al., "Growth inhibitors promote differentiation of insulin-producing tissue from embryonic stem cells," Proceedings of the National Academy of Sciences of the *United States of America 99: 16105-16110 (2002).*

Ivanova, N. B., *et al.,* "A Stem Cell Molecular Signature," Science 298: 601-604 (2002).

Jackson KA, Majka SM, Wang H, Pocius J, Hartley CJ, Majesky MW, Entman ML, Michael LH, Hirschi KK, Goodell MA (2001) Regeneration of ischemic cardiac muscle and vascular endothelium by adult stem cells. J Clin Invest 107:1395-1402.

Jaffredo T, Gautier R, Eichmann A, Dieterlen-Lievre F. Intraaortic hemopoietic cells are derived from endothelial cells during ontogeny. Development 1998;125:4575-4583.

James, Clive, "Global Review of Commercialized Transgenic Crops: 1998," *ISAAA*, No. 8, 1998.

Kahn, J., "Missing the mark on stem cells," *Bioethics Examiner* 5:3, Fall 2001, p. 1.

Kalka, C. et al. Transplantation of ex vivo expanded endothelial progenitor cells for therapeutic neovascularization. Proc. Natl. Acad. Sci. USA 2000; 97:3422-3427.

Kass, L., "The Meaning of Life - In the Laboratory," *The*

*Public Interest*, Winter 2002. Also see the Council's July 2002 report *Human Cloning and Human Dignity: An Ethical Inquiry,* Chapter 6.

Kawano H, Do YS, Kawano Y, Starnes V, Barr M, Law RE, Hsueh WA. Angiotensin II has multiple profibrotic effects in human cardiac fibroblasts. Circulation 2000;101:1130-1137.

Kennedy M, Firpo M, Choi K, Wall C, Robertson S, Kabrun N, Keller G. A common precursor for primitive erythropoiesis and definitive haematopoiesis.

Kennedy, D., "Stem Cells: Still Here, Still Waiting," *Science* 300: 865 (2003).

Kerr, D. A., et al., "*Hum*an Embryonic Germ Cell Derivatives Facilitate Motor Recovery of Rats with Diffuse Motor Neuron Injury," The Journal of Neuroscience 23: 5131-5140 (2003).

Kerr, D. A., *et al.,* "Human Embryonic Germ Cell Derivatives Facilitate Motor Recovery of Rats with Diffuse Motor Neuron Injury," The Journal of Neuroscience 23: 5131-5140 (2003).

Labastie M-C, Cortes F, Romeo P-H, Dulac C, Peault B. Molecular identity of hematopoietic precursor cells emerging in the human embryo. Blood 1998;92:3624-3635.

Latham, S., "Ethics and Politics," *American Journal of Bioethics* 2(1): 46 (2002).

Lechner, A. and Habener, J.F., "Stem/progenitor cells derived from adult tissues: potential for the treatment of diabetes mellitus," American Journal of Physiology—Endocrinology and Me*tabolism* 284: E259-266 (2003).

Lewis, C. S. The Abolition of Man. New York: The Macmillan Company, 1947.

Liker, M., et al., "Human neural stem cell transplantation in the MPTP-lesioned mouse," Brain Research 971: 168-177 (2003).

Liu, Z., and Mart*in, L. J.,* "Olfactory bulb core is a rich source of neural progenitor and stem cells in adult rodent and human," Journal·of Comparative Neurology 459: 368-391 (2003).

Liu, Z., and Mart*in,* L.J., "Olfactory bulb core is a rich source of neural progenitor and stem cells in adult rodent and human," Journal of Comparative Neurology 459: 368-391 (2003).

Los Angeles, CA.: University of California Press, 2003; also see Daniel Callahan's presentation before the Council on July 24, 2003, available on the Council's website at www.bioethics.gov.

Lumelsky, N., *et al.,* "Differentiation of embryonic stem cells to insulin-secreting structures similar to pancreatic islets," Science 292: 1389-1394 (2001).

Macklin, Ph.D., Ruth. "Ethics, Politics, and Human Embryo Stem Cell Research." Women's Health Issues, vol. 10, no. 3 (May/June 2000): 111-15.

Magnus, David, Arthur Caplan, and Glenn McGee, ed. In Who Owns Life? Amherst, New York: Prometheus Books, 2002.

Majumdar, M. K., et al., "Characterization and functionality of cell surface molecules on human mesenchymal stem cells," Journal of Biomedical Science 10: 228-241 (2003).

Marquis, D., "Stem cell research: The Failure of Bioethics," *Free Inquiry* 22(1): Winter 2002.

McCormick, R. A. "Who or What is the Preembryo?" In Corrective Vision: Explorations in Moral Theology. Kansas City: Sheed & Ward, 1994.

McEwan PE, Gray GA, Sherry L, Webb DJ, Kenyon CJ. Differential effects of angiotensin II on cardiac cell proliferation and intramyocardial perivascular fibrosis in vivo. Circulation 1998;98:2765-2773.

McGee, Glenn, and Elizabeth Banger. "Ethical Issues in the Patenting and Control of Stem Cell Research." In Who Owns Life? Ed. David Magnus, Arthur Caplan, and Glenn McGee. Amherst, New York: Prometheus Books, 2002. 243-264.

McKenny, Gerald P. "Enhancements and the Ethical Significance of Vulnerability." In Enhancing Human Traits: Ethical and Social Implications. Washington, D.C.: Georgetown University Press, 1998. 222-37.

McKibben, Bill. Enough: Staying Human in an Engineered Age. New York: Henry Holt and Company, 2003.

McLean, Margaret. "Stem Cells: Justice at the Gate." Dialog, vol. 41, no. 3 (fall 2002).

Meilaender, Gilbert. "Terra es animata: On Having a Life." Hastings Center Report, vol. 23, no. 4 (July-August 1993): 25-32.

Murray, T., "Hard Cell," *The American Prospect,* September 24, 2001.

Ogawa M, Kizumoto M, Nishikawa S, Fujimoto T, Kodama H, Nishikawa SI. Blood 1999;93:1168-1177.

Okarma, T., Presentation at the September 4, 2003, meeting of the President's Council on Bioethics, Washington, D.C., available at www.bioethics.gov.

Okarma, T., Presentation at the September 4, 2003, meeting of the President's Council on Bioethics, Washington, D.C., available at www.bioethics.gov.

Onyango, P., *et al.,* "Monoallelic expression and methylation of imprinted genes in human and mouse embryonic germ cell lineages," Proceedings of the National Academy of Sciences of the *United States of America* 99: 10599-10604 (2002).

Orlic D, Kajstura J, Chimenti S, et al. Bone marrow cells regenerate infarcted myocardium. Nature 2001, 410:701-705.

Orr, R., "The Moral Status of the Embryonal Stem Cell: Inherent or Imputed?" *American Journal of Bioethics* 2(1): 57-59 (2002).

Pagano, S.F., *et al.,* "Isolation and characterization of neural stem cells from the adult human olfactory bulb," Stem Cells 18: 295-300 (2000).

Pevny, L., and Rao, M.S., "The stem-cell menagerie," Trends in Neurosciences 26: 351-359 (2003).

Pitt B, Segal R, Martinez FA, et al. Randomised trial of losartan versus captopril in patients over 65 with heart failure (Evaluation of Losartan in the Elderly Study, ELITE). Lancet 1997;349:747-752.

Pittenger, M.F., *et al.,* "Adult mesenchymal stem cells: Potential for muscle and tendon regeneration and use in gene therapy," Journal of Musculoskeletal and Neuronal Interactions *2: 309-320 (2002).*

Pursley, W. H., Presentation at the September 4, 2003, meeting of the President's Council on Bioethics, Washington, D.C., available at www.bioethics.gov.

Pursley, W.H., Presentation at the September 4, 2003, meeting of the President's Council on Bioethics, Washington, D.C., available at www.bioethics.gov.

R. Alta Charo, of the University of Wisconsin Law School, quoted in Coyle, M., "The Clone Zone," *The National Law Journal*, May 15, 2002.

Rafii S, Shapiro F, Rimarachin J, Nachman R, Ferris B,

Weksler B, Moore AS, Asch AS. Isolation and characterization of human bone marrow microvascular endothelial cells: hematopoietic progenitor cell adhesion.

Rideout, W., *et al.*, "Correction of a genetic defect by nuclear transplantation and combined cell and gene therapy," Cell 109: 17-27 (2002).

Robertson, J., "Crossing the Ethical Chasm: Embryo Status and Moral Complicity," *American Journal of Bioethics* 2(1): 33 (2002).

Rohwedel, J., *et al.*, "Embryonic stem cells as an in vitro model for mutagenicity, cytotoxicity, and embryotoxicity studies: present state and future prospects," Toxicology In Vitro 15: 741-753 (2001).

Ryan, E.A., et al., "Clinical outcomes and insulin secretion after islet transplantation with the Edmonton protocol," Diabetes 50: 710-719 (2001).

Sato, N., *et al.*, "Molecular signature of human embryonic stem cells and its comparison with the mouse," Developmental Biology 260: 404-413 (2003).

Soria, B., *et al.*, "Insulin-secreting cells derived from embryonic stem cells normalize glycemia in streptozotocin-induced diabetic mice," Diabetes 49: 157-162 (2000)

Strong, C., "Those Divisive Stem Cells: Dealing with Our Most Contentious Issues," *American Journal of Bioethics* 2(1): 39-40 (2002).

Takahashi, T. et al. Ischemia- and cytokine-induced mobilization of bone marrowderived endothelial progenitor cells for neovascularization. Nat. Med. 1999; 5:434-438.

Tsai FY, Keller G, Kuo FC, Weiss M, Chen J, Rosenblatt M,

Alt FA, Orkin SH. An early hematopoietic defect in mice lacking the transcription factor GATA- 2. Nature 1994; 371:221-225.

Weissman, I. L., "Stem Cells—Scientific, Medical and Political Issues," *New England Journal of Medicine* 346(20): 1576-1579 (2002).

Xu, R. H., *et al.*, "BMP4 initiates human embryonic cell differentiation to trophoblast," Nature Biotechnology 20: 1261-1264 (2002).

Zalzman, M., *et al.*, "Reversal of hyperglycemia in mice using human expandable insulin-producing cells differentiated from fetal liver cells," Proceedings of the National Academy of Sciences of the *United States of America 100: 7253-7258 (2003).*

Zhang M, Methot D, Poppa V, Fujio Y, Walsh K, Murry CE (2001) Cardiomyocyte grafting for cardiac repair: graft cell death and anti-death strategies. J Mol Cell Cardiol 33:907-921.

Zhao, M., et al., "Amelioration of streptozotocin-induced diabetes in mice using human islet cells derived from long-term culture in vitro," Transplantation 73: 1454- 1460 (2002).

Zoloth, Laurie, "Jordon's Banks: A View from the First Years of Human Embryonic Stem Cell Research." In The Human Embryonic Stem Cell Debate. Ed. by Suzanne Holland, Karen Lebacqz, and Laurie Zoloth. Cambridge, MA: The MIT Press, 2001. 238.

Zulewski, H., *et al.*, "Multipotential Nestin-Positive Stem Cells Isolated From Adult Pancreatic Islets Differentiate Ex Vivo Into Pancreatic Endocrine, Exocrine and Hepatic Phenotypes," Diabetes 50: 521-533 (2001).

# Index


**A**

Abdel-Fattah, 213
Adonai, 137
Alfred Russel Wallace, 204
Angiosperms, 198
ANOVA, 218, 219
Anoxygenic, 194
Anunnaki, 120, 128
  genes, 142
  skill level, 139
APHA, 228
Arundel, 225
AWWA, 228

**B**

Banerjee, 225
Basic Local Alignment Search Tool, 105
Beneficial alleles, 168
Biofilms, 224
Bioinformatic Harvester, 114
Bioinformatics, 102
  Applications, 104
  Tools, 105
Biotechnology, 102
  Software Tools, 114
Bishop James, 123
BLAST, 105

**C**

C.D. Darlington, 206
Cauliflower Mosaic Virus, 209
CBR, 104
Charles Darwin, 143, 172
Cloete, 224
COD, 223
Computer software, 102
CSIRO, 96
Culture Media, 215

**D**

Darwin, 118
Darwinian,
  machine, 108
  models, 125
Decrease Genetic Variation, 150
Deleterious alleles, 167
Devonian period, 197
DNA, 99, 170, 178, 192, 206
  forces, 206
  sequences, 107
Duta, 213

## E

EMBOSS, 115
Entrez, 114
Enuma Elish document, 126
Ernst Mayr, 206
EST, 109
Evolution Among Lineages, 177
Evolutionary,
  Biology, 117
  Theory, 174

## F

FACT, 116
Futuyma, D.J., 117

## G

G.G. Simpson, 206
G. Ledyard Stebbins, 206
G. Magnirostris, 156
Gene,
  Expression Omnibus, 263
  Flow, 170
  Gechnology, 95
  Technology Act 2000, 98
  Technology, 98
GeneMap98, 106
Genetic Drift, 162
Genetic,
  engineering, 102
  evidence, 133
  exploration, 119
  variation, 164
GEO, 262
George C. Williams, 207
Glutaraldehyde, 224
GMO, 97
Grail Kings, 123

## H

H.B.D. Kettlewell, 144
Harrington, 131
Harvard University, 123
Holy Grail, 123
Homo Erectus, 139, 142
  genes, 139
HPLC, 216
HT, 110
Human Genome, 211
Hybridizations, 261

## I

Insertion package, 207

## J

J.B.S. Haldane, 153, 175
John Maynard Smith, 207
Judaism, 137

## K

Kapadnis, 213
Karl Pearson, 204
Kim, 213

## L

Lazarova, 224
Lodish, 143

## M

MAGEML, 262
MAML, 262, 263
Manem, 224
Markov Chain Monte Carlo, 112
Mass Extinction, 189
Mendel's Laws, 174
Mendelian school, 205
MGED, 263
Microarray Gene Expression Database, 263
Microarray Markup Language, 263
Microbial biofilms, 224
Microorganism, 215
Modern,
  Evolutionary Synthesis, 203
  Synthesis, 117, 206
Modes of Speciation, 186
Molecular diagnostics, 263
Montgomery, 219
MPSS, 109
Mr. Ventair, 211
MUSCLE, 115
Mutation, 164

## N

Natural Selection, 116, 150, 160
NCBI, 263
NCBIO, 104
Neo-Darwinian, 119
New Paradigm for Biology, 207
NRC, 104

## O

Olama, 214
Old Testament, 136
OMIM, 106
Online Mendelian Inheritance in Man, 106
Ordinary Extinction, 188
Orthology analysis, 112
Owen White, 107

## P

Permian extinction, 198
Photosynthesis, 193
Planet X, 120

## R

R.A. Fisher, 175, 205
RBCs, 223
RNA, 178
  variation, 263
  world hypothesis, 192
Roman Church, 137
Russell, 224

## S

SAGE, 109
Scientific creationism, 200
Sewall Wright, 175
Sexual Selection, 160
Sir Laurence Gardner, 123
Sofer, 224
Species barrier, 209
STING, 114

### T

T.H. Morgan, 205
Tailoring crops, 98
Tanyildizi, 214
Thomas Kuhn, 124
Trait Gene, 208, 209

### U

Unpredictable, 210

### V

Van Flandern, 131

### W

W.D. Hamilton, 207
Walter Frank Raphael Weldon, 204
WPCE, 228

### Z

Zecharia Sitchin, 119